Research Notes in Mathematics

Submission of proposals for consideration
Suggestions for publication, in the form of outlines and representative samples, are invited by the editorial board for assessment. Intending authors should contact either the main editor or another member of the editorial board, citing the relevant AMS subject classifications. Refereeing is by members of the board and other mathematical authorities in the topic concerned, located throughout the world.

Preparation of accepted manuscripts
On acceptance of a proposal, the publisher will supply full instructions for the preparation of manuscripts in a form suitable for direct photo-lithographic reproduction. Specially printed grid sheets are provided and a contribution is offered by the publisher towards the cost of typing.

Illustrations should be prepared by the authors, ready for direct reproduction without further improvement. The use of hand-drawn symbols should be avoided wherever possible, in order to maintain maximum clarity of the text.

The publisher will be pleased to give any guidance necessary during the preparation of a typescript, and will be happy to answer any queries.

Important note
In order to avoid later retyping, intending authors are strongly urged not to begin final preparation of a typescript before receiving the publisher's guidelines and special paper. In this way it is hoped to preserve the uniform appearance of the series.

Titles in this series

1 Improperly posed boundary value problems
A Carasso and A P Stone
2 Lie algebras generated by finite dimensional ideals
I N Stewart
3 Bifurcation problems in nonlinear elasticity
R W Dickey
4 Partial differential equations in the complex domain
D L Colton
5 Quasilinear hyperbolic systems and waves
A Jeffrey
6 Solution of boundary value problems by the method of integral operators
D L Colton
7 Taylor expansions and catastrophes
T Poston and I N Stewart
8 Function theoretic methods in differential equations
R P Gilbert and R J Weinacht
9 Differential topology with a view to applications
D R J Chillingworth
10 Characteristic classes of foliations
H V Pittie
11 Stochastic integration and generalized martingales
A U Kussmaul
12 Zeta-functions: An introduction to algebraic geometry
A D Thomas
13 Explicit *a priori* inequalities with applications to boundary value problems
V G Sigillito
14 Nonlinear diffusion
W E Fitzgibbon III and H F Walker
15 Unsolved problems concerning lattice points
J Hammer
16 Edge-colourings of graphs
S Fiorini and R J Wilson
17 Nonlinear analysis and mechanics: Heriot-Watt Symposium Volume I
R J Knops
18 Actions of fine abelian groups
C Kosniowski
19 Closed graph theorems and webbed spaces
M De Wilde
20 Singular perturbation techniques applied to integro-differential equations
H Grabmüller
21 Retarded functional differential equations: A global point of view
S E A Mohammed
22 Multiparameter spectral theory in Hilbert space
B D Sleeman
24 Mathematical modelling techniques
R Aris
25 Singular points of smooth mappings
C G Gibson
26 Nonlinear evolution equations solvable by the spectral transform
F Calogero
27 Nonlinear analysis and mechanics: Heriot-Watt Symposium Volume II
R J Knops
28 Constructive functional analysis
D S Bridges
29 Elongational flows: Aspects of the behaviour of model elasticoviscous fluids
C J S Petrie
30 Nonlinear analysis and mechanics: Heriot-Watt Symposium Volume III
R J Knops
31 Fractional calculus and integral transforms of generalized functions
A C McBride
32 Complex manifold techniques in theoretical physics
D E Lerner and P D Sommers
33 Hilbert's third problem: scissors congruence
C-H Sah
34 Graph theory and combinatorics
R J Wilson
35 The Tricomi equation with applications to the theory of plane transonic flow
A R Manwell
36 Abstract differential equations
S D Zaidman
37 Advances in twistor theory
L P Hughston and R S Ward
38 Operator theory and functional analysis
I Erdelyi
39 Nonlinear analysis and mechanics: Heriot-Watt Symposium Volume IV
R J Knops
40 Singular systems of differential equations
S L Campbell
41 N-dimensional crystallography
R L E Schwarzenberger
42 Nonlinear partial differential equations in physical problems
D Graffi
43 Shifts and periodicity for right invertible operators
D Przeworska-Rolewicz
44 Rings with chain conditions
A W Chatters and C R Hajarnavis
45 Moduli, deformations and classifications of compact complex manifolds
D Sundararaman
46 Nonlinear problems of analysis in geometry and mechanics
M Atteia, D Bancel and I Gumowski
47 Algorithmic methods in optimal control
W A Gruver and E Sachs
48 Abstract Cauchy problems and functional differential equations
F Kappel and W Schappacher
49 Sequence spaces
W H Ruckle
50 Recent contributions to nonlinear partial differential equations
H Berestycki and H Brezis
51 Subnormal operators
J B Conway
52 Wave propagation in viscoelastic media
F Mainardi
53 Nonlinear partial differential equations and their applications: Collège de France Seminar. Volume I
H Brezis and J L Lions
54 Geometry of Coxeter groups
H Hiller
55 Cusps of Gauss mappings
T Banchoff, T Gaffney and C McCrory

56 An approach to algebraic K-theory
A J Berrick
57 Convex analysis and optimization
J-P Aubin and R B Vintner
58 Convex analysis with applications in the differentiation of convex functions
J R Giles
59 Weak and variational methods for moving boundary problems
C M Elliott and J R Ockendon
60 Nonlinear partial differential equations and their applications: Collège de France Seminar. Volume II
H Brezis and J L Lions
61 Singular systems of differential equations II
S L Campbell
62 Rates of convergence in the central limit theorem
Peter Hall
63 Solution of differential equations by means of one-parameter groups
J M Hill
64 Hankel operators on Hilbert space
S C Power
65 Schrödinger-type operators with continuous spectra
M S P Eastham and H Kalf
66 Recent applications of generalized inverses
S L Campbell
67 Riesz and Fredholm theory in Banach algebra
B A Barnes, G J Murphy, M R F Smyth and T T West
68 Evolution equations and their applications
F Kappel and W Schappacher
69 Generalized solutions of Hamilton-Jacobi equations
P L Lions
70 Nonlinear partial differential equations and their applications: Collège de France Seminar. Volume III
H Brezis and J L Lions
71 Spectral theory and wave operators for the Schrödinger equation
A M Berthier
72 Approximation of Hilbert space operators I
D A Herrero
73 Vector valued Nevanlinna Theory
H J W Ziegler
74 Instability, nonexistence and weighted energy methods in fluid dynamics and related theories
B Straughan
75 Local bifurcation and symmetry
A Vanderbauwhede
76 Clifford analysis
F Brackx, R Delanghe and F Sommen
77 Nonlinear equivalence, reduction of PDEs to ODEs and fast convergent numerical methods
E E Rosinger
78 Free boundary problems, theory and applications. Volume I
A Fasano and M Primicerio
79 Free boundary problems, theory and applications. Volume II
A Fasano and M Primicerio
80 Symplectic geometry
A Crumeyrolle and J Grifone
81 An algorithmic analysis of a communication model with retransmission of flawed messages
D M Lucantoni
82 Geometric games and their applications
W H Ruckle
83 Additive groups of rings
S Feigelstock
84 Nonlinear partial differential equations and their applications: Collège de France Seminar. Volume IV
H Brezis and J L Lions
85 Multiplicative functionals on topological algebras
T Husain
86 Hamilton-Jacobi equations in Hilbert spaces
V Barbu and G Da Prato
87 Harmonic maps with symmetry, harmonic morphisms and deformations of metrics
P Baird
88 Similarity solutions of nonlinear partial differential equations
L Dresner
89 Contributions to nonlinear partial differential equations
C Bardos, A Damlamian, J I Díaz and J Hernández
90 Banach and Hilbert spaces of vector-valued functions
J Burbea and P Masani
91 Control and observation of neutral systems
D Salamon
92 Banach bundles, Banach modules and automorphisms of C*-algebras
M J Dupré and R M Gillette
93 Nonlinear partial differential equations and their applications: Collège de France Seminar. Volume V
H Brezis and J L Lions

Control and observation of neutral systems

D Salamon

University of Bremen

Control and observation of neutral systems

Pitman Advanced Publishing Program

BOSTON · LONDON · MELBOURNE

PITMAN PUBLISHING LIMITED
128 Long Acre, London WC2E 9AN

PITMAN PUBLISHING INC
1020 Plain Street, Marshfield, Massachusetts 02050

Associated Companies
Pitman Publishing Pty Ltd, Melbourne
Pitman Publishing New Zealand Ltd, Wellington
Copp Clark Pitman, Toronto

First published 1984

AMS Subject Classifications: (main) 34K05, 34K35, 93D15
(subsidiary) 45E10, 47A55

Library of Congress Cataloging in Publication Data

Salamon, D. (Dietmar)
Control and observation of neutral systems.

(Research notes in mathematics; 91)
Bibliography: p.
1. Functional differential equations—Delay equations.
2. Control theory. I. Title. II. Series.
QA372.S15 1984 515.3′52 83-13322
ISBN 0-273-08618-9

British Library Cataloguing in Publication Data

Salamon, Dietmar
Control and observation of neutral systems.—
(Research notes in mathematics; 91)
1. Volterra equations
I. Title. II. Series
515′.45 QA431

ISBN 0-273-08618-9

Reproduced and printed by photolithography
in Great Britain by Biddles Ltd, Guildford

Contents

The author wishes to express his sincere thanks to Professor Dr D. Hinrichsen (University of Bremen) for his critical interest in this research work and many stimulating discussions.

Preface

In this research note a theory of control and observation is presented for linear neutral functional differential equations (NFDE) with general delays in the state and input/output variables. We consider the controlled NFDE

$$\frac{d}{dt}\left(x(t) - Mx_t - \Gamma u_t\right) = Lx_t + Bu_t \tag{1}$$

and the observed NFDE

$$\dot{x}(t) = L^T x_t + M^T \dot{x}_t, \quad y(t) = B^T x_t + \Gamma^T \dot{x}_t, \tag{2}$$

which is obtained from (1) by transposition of matrices. For systems of this form three types of problem, namely, completeness and small solutions (Chapter 3), controllability and observability (Chapter 4), state feedback and dynamic observation (Chapter 5), are discussed within the context of functional analytic semigroup theory. Functional differential equations (FDE) have been studied in this context for about 20 years. The present work is mainly influenced by three recent developments in this area which have made the linear theory much more elegant and efficient.

The first is the introduction of so-called structural operators for the state space description of retarded functional differential equations (RFDE) in the product space $\mathbb{R}^n \times L^p$. The basic ideas in this direction were presented in 1976 in a paper by Bernier and Manitius [11] and further developments can be found, for example, in Manitius [93] and Delfour-Manitius [29]. In particular, these results have been applied to problems of completeness and approximate controllability (Manitius [93-95]). The concept of structural operators has been extended to retarded systems with input delays by Vinter-Kwong [147] and Delfour [28]. This had led to an evolution equation for the state space description of RFDEs with input delays.

A second important development took place within the duality theory of RFDEs. It is well known that the adjoint of a semigroup associated with a delay equation is not of the same type as the original one; however, only fairly recently has the adjoint semigroup been interpreted in terms of the underlying system equation. Such an interpretation was first given by Burns and Herdman [17] for Volterra integro-differential equations. These authors show that the adjoint semigroup is associated with the transposed equation via an alternative state concept which is due to Miller [104]. Further results in this direction can be found, for example, in Diekmann [32, 33]. This duality theory via the two notions of the state is closely related to the concept of the structural operators: in fact, the structural operators describe the relation between the two state concepts.

A third development is an extension to neutral systems of the semigroup approach in the product space $\mathbb{R}^n \times L^p$. This has been presented in two recent papers by Burns, Herdman and Stech [18, 19].

The theory of NFDEs in the product space framework is not very far developed at present. In particular, a satisfactory duality theory is still missing and structural operators for the description of neutral systems have not yet been introduced. The latter has been stated as an open problem in Delfour [28, Remark 2.5]. In Chapter 2 this gap is filled. However, a straighforward extension of the results on retarded systems is not possible. The two state concepts for RFDEs have both been described in the same space $\mathbb{R}^n \times L^p$. In the case of neutral systems we are forced to work in the two state spaces $W^{1,p}$ and $\mathbb{R}^n \times L^p$, where $W^{1,p}$ is embedded into $\mathbb{R}^n \times L^p$ as a dense subspace (see Section 2.1).

Almost half of the book is devoted to the development of this state space approach (duality, structural operators, evolution equations, spectral theory). These results provide the framework for the treatment of the structural properties of neutral systems (Chapter 3) as well as the control and observation problems (Chapters 4 and 5). The main results in Chapters 3, 4 and 5 are the following:

- duality between completeness and nonexistence of nonzero small solutions (Theorem 3.1.10);
- duality between F-completeness and nonexistence of nontrivial small solutions (Theorem 3.2.3);
- a matrix-type condition for F-completeness (Corollary 3.2.5);
- a characterization of spectral controlability and observability (Proposition 4.1.2 and Theorem 4.1.11);
- duality between approximate controllability and strict observability (Theorem 4.2.6);

- a matrix-type condition for observability of small solutions (Theorem 4.2.11);
- duality between approximate F-controlability and observability (Theorem 4.3.5);
- a matrix-type condition for observability of nontrivial small solutions (Theorem 4.3.7);
- a concrete representation of the observer semigroup (Theorem 5.2.1);
- the 'spectrum-determined growth' property of the observer semigroup (Theorem 5.2.7);
- finite pole-shifting (Theorem 5.3.2).

A preliminary chapter discusses some basic facts concerning Volterra-Stieltjes integral equations (Section 1.1) and systems of functional and functional differential equations (Section 1.2). Moreover a general framework is presented for the study of infinite-dimensional linear systems with unbounded input/output operators (Section 1.3). The essential point in this section is that the semigroup is not assumed to have any smoothing property. Such results are needed for the treatment of retarded and neutral systems with point delays in input and output.

1 Preliminaries

1.1 VOLTERRA-STIELTJES INTEGRAL EQUATIONS

In this section we deal with existence, uniqueness, continuous dependence and representation of L^p-solutions to the Volterra-Stieltjes integral equation

$$x(t) = \int_0^t d\alpha(s)x(t-s) + f(t), \quad 0 \leqslant t < T, \tag{1}$$

where $f \in L^p([0,T];\mathbb{R}^n)$ and $\alpha \in NBV([0,T);\mathbb{R}^{n\times n})$, i.e., $\alpha(t)$ is of bounded variation on $[0,T)$, right continuous for $0 < t < T$, and $\alpha(0) = 0$.

We cannot expect existence and uniqueness for the solutions of (1) with an arbitrary α. For example, if $\alpha = \rho \in NBV([0,T);\mathbb{R}^{n\times n})$ is given by $\rho(0) = 0$ and $\rho(t) = I$ for $0 < t < T$, then (1) is equivàlent to $f(t) \equiv 0$. To exclude such a situation, we will always assume that

$$1 \notin \sigma(A_0) \tag{2}$$

where $A_0 \in \mathbb{R}^{n\times n}$ is given by $A_0 = \lim_{t\downarrow 0} \alpha(t)$. Note that the latter implies

$$\lim_{t\downarrow 0} \operatorname*{VAR}_{[0,t]} [\alpha - A_0\rho] = 0. \tag{3}$$

Now let us assemble some basic facts on the convolution of measures and functions.

1.1.1 REMARKS

(i) Let $1 < p \leqslant \infty$, $1/p + 1/q = 1$, and $g \in L^q[0,T]$. Then the function

$$g * f(t) = \int_0^t g(s)f(t-s)ds, \quad 0 \leqslant t \leqslant T,$$

is continuous for every $f \in L^p[0,T]$ (see, e.g., Hewitt-Ross [52, Theorem 20.16]). Moreover the convolution operator $f \to g * f$ from $L^p[0,T]$ into $C[0,T]$ is compact. This follows from the Arzela-Ascoli theorem and the inequality

$$|g * f(t) - g * f(s)| \leq \|f\|_p \left(\int_0^t |g(t-\tau) - g(s-\tau)|^q d\tau\right)^{1/q}$$

for $0 \leq s \leq t \leq T$ (define $g(t) := 0$ for $t \notin [0,T]$).

(ii) Let $f \in L^p[0,T]$, $1 \leq p < \infty$, and $g \in L^1[0,T]$. Then $g * f \in L^p[0,T]$ and

$$\|g * f\|_p \leq \|g\|_1 \ \|f\|_p \tag{4}$$

(see, e.g., Hewitt-Ross [52, Corollary 20.14]). Moreover the operator $f \to g * f$ on $L^p[0,T]$ is compact. In fact, for all $\varepsilon > 0$ and almost all $t \in [-\varepsilon,T]$ we have

$$|g * f(t+\varepsilon) - g * f(t)| \leq \int_{-\varepsilon}^t |g(s+\varepsilon)-g(s)||f(t-s)|ds = \tilde{g} * \tilde{f}(t)$$

where $\tilde{g}(t) := |g(t+\varepsilon) - g(t)|$ for $-\varepsilon \leq t \leq T$ and $\tilde{f}(t) := |f(t)|$ for $0 \leq t \leq T$ (again $g(t) := 0$ and $f(t) := 0$ for $t \notin [0,T]$). Applying (4) to the right-hand side of the above inequality, we obtain

$$\left(\int_{-\varepsilon}^T |g * f(t+\varepsilon) - g * f(t)|^p dt\right)^{1/p} \leq \|f\|_p \int_{-\varepsilon}^T |g(t+\varepsilon) - g(t)|dt.$$

A similar inequality holds for $\varepsilon < 0$. Hence compactness follows from the analogue of the Arzela-Ascoli theorem for L^p spaces which is due to M. Riesz (see e.g., Dunford and Schwartz [37, Theorem 4.8.20]).

(iii) Every $\alpha \in NBV[0,T)$ represents a Borel measure on $\mathbb{R}$ with no mass outside the interval $[0,T)$. This measure will be denoted by $d\alpha$. For any $d\alpha$-integrable function $g : [0,T] \to \mathbb{R}$ the expression

$$d\alpha(g) = \int_0^T g(t)d\alpha(t) \tag{5}$$

denotes the integral of g with respect to the measure $d\alpha$. If g is continuous, then (5) can be understood as a Stieltjes integral.

(iv) By the Riesz representation theorem, $NBV[0,T)$ is (isometrically isomorphic to) the dual space X_T^* of

$$X_T = \{g \in C[0,T] \mid g(T) = 0\}.$$

(v) For $\alpha, \beta \in NBV[0,T)$ let $d\alpha * d\beta$ be the convolution of the Borel measures $d\alpha$ and $d\beta$ restricted to the interval $[0,T)$. This means that $d\alpha * d\beta$ is given by the relation

$$\begin{aligned}[d\alpha * d\beta](g) &= \int_0^T \int_0^{T-s} g(t+s)d\beta(t)d\alpha(s) \\ &= \int_0^T \int_0^{T-t} g(t+s)d\alpha(s)d\beta(t)\end{aligned} \tag{6}$$

for every $g \in X_T$ (Hewitt-Ross [52, Chapter 19]).

(vi) For $\alpha \in NBV[0,T)$ and $f \in L^p_{loc}(\mathbb{R})$, $1 \leq p \leq \infty$, let $d\alpha * f \in L^p_{loc}(\mathbb{R})$ be the convolution of the Borel measure $d\alpha$ and the function f. In the case $p > 1$, this means that $d\alpha * f$ is defined by the relation

$$\int_a^b g(t)[d\alpha * f](t)dt = \int_0^T \int_{a-s}^{b-s} g(t+s)f(t)dt\, d\alpha(s) \tag{7}$$

for all $a, b \in \mathbb{R}$, $a \leq b$, and $g \in L^q[a,b]$, $1/p + 1/q = 1$, (Hewitt-Ross [52, Definition 20.5]).

Moreover the following inequality holds for every $f \in L^p[a,b]$

$$\|d\alpha * f\|_p \leq \underset{[0,T)}{\mathrm{VAR}}\, \alpha \, \|f\|_p \tag{8}$$

(Hewitt-Ross [52, Theorem 20.12]).

(vii) Let $\alpha \in NBV[0,T)$ and $f \in L^p[0,T]$, $1 \leq p \leq \infty$. Then $d\alpha * f \in L^p[0,T]$ can also be defined by the explicit expression

$$[d\alpha * f](t) = \int_0^t d\alpha(s)f(t-s)$$

for almost every $t \in [0,T]$ (Hewitt-Ross [52, Theorem 20.9]).

(viii) Let $\alpha, \beta \in NBV[0,T)$. Then $d\alpha * \beta = \alpha * d\beta \in NBV[0,T)$ and

$$d[d\alpha * \beta] = d\alpha * d\beta. \tag{9}$$

Proof. Let $\gamma \in NBV[0,T)$ be chosen such that $d\gamma = d\alpha * d\beta$. Then, for every $g \in L^q[0,T]$, $1 \leqslant q < \infty$, the following equation holds

$$\int_0^T g(t)\gamma(t)dt = \int_0^T \int_t^T g(\tau)d\tau d\gamma(t)$$

$$= \int_0^T \int_0^{T-s} \int_{t+s}^T g(\tau)d\tau d\beta(t)d\alpha(s), \text{ by (6)},$$

$$= \int_s^T \int_0^{T-s} g(t+s)\beta(t)dtd\alpha(s)$$

$$= \int_0^T g(t)[d\alpha * \beta](t)dt, \text{ by (7)}.$$

Hence $\gamma(t) = d\alpha * \beta(t)$ for almost all $t \in [0,T)$. This proves the statement.

(ix) Note that the Borel measure $d\alpha$ can also be interpreted as the distributional derivative of $\alpha \in NBV[0,T)$. In this sense (9) follows from the fact that - in order to differentiate a convolution product of distributions - it suffices to differentiate one of the factors.

(x) Let $\alpha \in NBV[0,T)$, $\beta \in W^{1,p}[0,T]$, and $f \in L^p[0,T]$ such that $\beta(0) = 0$ and $\dot{\beta} = f$. Moreover let $\rho \in NBV[0,T)$ be defined by $\rho(0) = 0$ and $\rho(t) = 1$ for $t > 0$. Then

$$d\alpha * \beta = d\alpha * (\rho * \dot{\beta}) = \alpha * f = \rho * (d\alpha * f).$$

Hence $d\alpha * \beta = \alpha * f$ is absolutely continuous with derivative $d\alpha * f \in L^p[0,T]$ and satisfies $d\alpha * \beta(0) = 0$.

(xi) In the vectorial case the above definitions have to be understood componentwise. In particular, for $\alpha \in NBV([0,T);\mathbb{R}^{m\times n})$ and $f \in L^p([0,T];\mathbb{R}^n)$, $1 < p \leqslant \infty$, the function $d\alpha * f \in L^p([0,T];\mathbb{R}^m)$ is defined by the equation

$$\int_0^T g^T(t)[d\alpha * f](t)dt = \int_0^T \int_s^T g^T(t)d\alpha(s)f(t-s)dt$$

$$= \sum_{i=1}^m \sum_{j=1}^n \int_0^T \int_s^T g_i(t)f_j(t-s)dtd\alpha_{ij}(s)$$

for all $g \in L^q([0,T];\mathbb{R}^m)$, $1/p + 1/q = 1$ (in the second term α and f cannot be interchanged due to the matrix notation).

1.1.2 THEOREM. *Let* $\alpha \in NBV([0,T);\mathbb{R}^{n\times n})$ *satisfy* (2). *Then*

(i) *for every* $f \in C([0,T];\mathbb{R}^n)$ *with* $f(0) = 0$ *there exists a unique solution* $x \in C([0,T];\mathbb{R}^n)$ *of* (1) *satisfying* $x(0) = 0$ *and depending continuously on* f *with respect to the sup-norm;*

(ii) *for every* $f \in NBV([0,T);\mathbb{R}^n)$ *there exists a unique solution* $x \in NBV([0,T);\mathbb{R}^n)$ *of* (1), *i.e.* $x = d\alpha * x + f$, *depending continuously on* f *with respect to the* NBV-*norm.*

Proof. (i) Let X denote the Banach space of all $x \in C([0,T];\mathbb{R}^n)$ with $x(0) = 0$, endowed with the sup-norm. Then the function

$$[Ax](t) = d\alpha * x(t) = \int_0^t d\alpha(s)x(t-s), \ 0 \leq t \leq T,$$

is in X for every $x \in X$. Moreover the linear operator $A : X \to X$ is bounded. We have to show that $1 \notin \sigma(A)$.

For this purpose let $\varepsilon > 0$ and $\gamma > 0$ be chosen such that

$$\underset{[0,\varepsilon]}{\mathrm{VAR}} [\alpha - A_0\rho] + e^{-\gamma\varepsilon} \underset{[0,T)}{\mathrm{VAR}} [\alpha - A_0\rho] < |(I-A_0)^{-1}|^{-1}$$

and define on X the equivalent norm

$$\|x\|_\gamma = \sup_{0\leq t\leq T} |x(t)|e^{-\gamma t}, \ x \in X.$$

Then we have the following estimation for every $x \in X$ and $t \in [0,T]$

$$|[Ax](t) - A_0x(t)|e^{-\gamma t}$$

$$\leq |\int_0^\varepsilon d[\alpha - A_0\rho](s)x(t-s) \ e^{-\gamma t} + |\int_\varepsilon^t d[\alpha - A_0\rho](s)x(t-s)|e^{-\gamma t}$$

$$\leq \underset{[0,\varepsilon]}{\mathrm{VAR}} [\alpha - A_0\rho] \sup_{t-\varepsilon\leq s\leq t} |x(s)|e^{-\gamma t}$$

$$+ \underset{[\varepsilon,t)}{\mathrm{VAR}} [\alpha - A_0\rho] \sup_{0\leq s\leq t-\varepsilon} |x(s)|e^{-\gamma t}$$

$$\leq \left(\underset{[0,\varepsilon]}{\mathrm{VAR}} [\alpha - A_0\rho] + e^{-\gamma\varepsilon} \underset{[0,T)}{\mathrm{VAR}} [\alpha - A_0\rho] \right) \|x\|_\gamma .$$

Hence, by definition of ε and γ, the affine map

$$x \to (I-A_0)^{-1}[Ax - A_0x + f]$$

is a contraction on X with respect to $\| \cdot \|_\gamma$. The unique fixed point of this map is precisely the solution of $[I - A]x = f$.

(ii) The dual space X^* of X can be represented by $NBV = NBV([0,T);\mathbb{R}^n)$ via the pairing

$$\langle g,\beta\rangle = \int_0^T g^T(T-t)d\beta(t), \; g \in X, \quad \beta \in NBV,$$

(compare Remark 1.1.1 (iv)). Moreover, the following equation holds

$$\langle Ag,\beta\rangle = \int_0^T \left(\int_0^{T-t} d\alpha(s)g(T-t-s)\right)^T d\beta(t)$$

$$= \int_0^T \int_0^{T-t} g^T(T-t-s)d\alpha^T(s)d\beta(t)$$

$$= \int_0^T g^T(T-t)d(d\alpha^T * \beta)(t)$$

$$= \langle g, d\alpha^T * \beta\rangle, \; g \in X, \quad \beta \in NBV,$$

where the last but one equality follows from (9) and (6). We conclude that the adjoint operator A^* on $X^* \approx NBV$ is given by

$$A^*\beta = d\alpha^T * \beta, \quad \beta \in NBV.$$

Hence statement (ii) - applied to α^T - follows from (i) and the fact that $\sigma(A^*) = \sigma(A)$. □

1.1.3 <u>DEFINITION</u>. *Let* $\alpha \in NBV([0,T);\mathbb{R}^{n\times n})$ *satisfy* (2). *Then the unique solution* $\xi \in NBV([0,T);\mathbb{R}^{n\times n})$ *of*

$$\xi = d\alpha * \xi + \rho \tag{10}$$

($\rho(0) = 0$, $\rho(t) = I$ *for* $t > 0$) *is said to be the fundamental solution of* (1).

If $\alpha \in NBV([0,T);\mathbb{R}^{n\times n})$ satisfies (2) and ξ is the fundamental solution of (1), then

$$\xi = \xi * d\alpha + \rho. \tag{11}$$

In fact, the unique solution $\zeta \in NBV([0,T);\mathbb{R}^{n\times n})$ of (11) satisfies

$$\begin{aligned}\zeta &= \zeta * d\rho = \zeta * d[\xi - d\alpha * \xi] \\ &= d[\zeta - \zeta * d\alpha] * \xi = d\rho * \xi = \xi.\end{aligned}$$

Note that, by (11), ξ^T is the fundamental solution of the transposed equation, $x = d\alpha^T * x + f$.

Now we are in a position to prove the main result of this section.

1.1.4 THEOREM. *Let* $\alpha \in NBV([0,T);\mathbb{R}^{n\times n})$ *satisfy* (2) *and let* ξ *be the fundamental solution of* (1). *Then, for every* $f \in L^p([0,T];\mathbb{R}^n)$, *there exists a unique solution* $x \in L^p([0,T];\mathbb{R}^n)$ *of* (1), *i.e.*, $x = d\alpha * x + f$. *This solution is given by*

$$x = d\xi * f \tag{12}$$

and depends continuously on f *with respect to the* L^p*-norm.*

Proof. Let $x \in L^p([0,T];\mathbb{R}^n)$ be given by (12). Then, by (10),

$$\begin{aligned}x &= d\xi * f = d(d\alpha * \xi + \rho) * f \\ &= d\alpha * (d\xi * f) + d\rho * f = d\alpha * x + f.\end{aligned}$$

Conversely, let x be a solution of (1). Then, by (11),

$$\begin{aligned}x &= d\rho * x = d(\xi - \xi * d\alpha) * x \\ &= d\xi * (x - d\alpha * x) = d\xi * f.\end{aligned}$$

This proves the existence, the uniqueness and (12). Continuous dependence follows from (12) and (8). □

Our final result follows directly from Theorem 1.1.4 and Remark 1.1.1 (x).

1.1.5 <u>COROLLARY</u>. *Let* $\alpha \in NBV([0,T);\mathbb{R}^{n\times n})$ *satisfy* (2), $f \in L^p([0,T];\mathbb{R}^n)$, *and let* $x \in L^p([0,T];\mathbb{R}^n)$ *be the unique solution of* (1). *Then* $f \in W^{1,p}([0,T];\mathbb{R}^n)$ *and* $f(0) = 0$ *if and only if* $x \in W^{1,p}([0,T];\mathbb{R}^n)$ *and* $x(0) = 0$. *Moreover, in this case* $\dot{x} = \dot{f} + d\alpha * \dot{x}$.

REMARKS ON THE LITERATURE

In the theory of Volterra-Stieltjes integral equations the central result is the existence and uniqueness of the resolvent kernel (in our notation the fundamental solution). This has first been proved by Hinton [54], Bitzer [14], and later on with different methods by Schwabik [137, 138]. Kappel [67] has shown the existence and uniqueness of the fundamental solution via Laplace transform methods. Extensions to infinite-dimensional spaces can be found in Hönig [55, 56].

Note that the existence of the fundamental solution is the only assumption which is needed for the proof of Theorem 1.1.4, and that Theorem 1.1.2 can be regarded as a corollary of formula (12) together with Remark 1.1.1. An alternative proof of Theorem 1.1.2 is given in order to make this work self-contained.

Moreover, note that none of the references mentioned above considers solutions of (1) in the function space $L^p([0,T];\mathbb{R}^n)$. This has only been done by Burns, Herdman and Stech [19]. However, the existence and uniqueness result in [19, Lemma 2.6] - when applied to equation (1) - leads only to inhomogeneous terms f of a special form.

Burns, Herdman and Stech [19, Section 4] have also shown by an example that condition (2) is not necessary in order to prove existence and uniqueness for the L^p-solutions of (1). However, it is known (Schwabik [137, 138], Hönig [55, 56]) that (2) is necessary and sufficient for the existence and uniqueness of the fundamental solution.

1.2 SYSTEMS OF FUNCTIONAL AND FUNCTIONAL DIFFERENTIAL EQUATIONS

In this section we develop some basic results on systems of the form

$$\begin{aligned} \dot{w}(t) &= Lw_t + Bx_t + f(t) \\ x(t) &= \Gamma w_t + Mx_t + g(t) \end{aligned} \tag{13}$$

where $w(t) \in \mathbb{R}^n$, $x(t) \in \mathbb{R}^m$, $w_t(\tau) = w(t+\tau)$ for $-a \leq \tau \leq 0$, $x_t(\tau) = x(t+\tau)$ for $-h \leq \tau \leq 0$, and L, B, Γ, M are bounded linear functionals on the appropriate spaces of continuous functions, given by

$$L\phi = \int_{-a}^{0} d\eta(\tau)\phi(\tau), \quad \phi \in C([-a,0];\mathbb{R}^n),$$

$$B\phi = \int_{-h}^{0} d\beta(\tau)\phi(\tau), \quad \phi \in C([-h,0];\mathbb{R}^m),$$

$$\Gamma\phi = \int_{-a}^{0} d\gamma(\tau)\phi(\tau), \quad \phi \in C([-a,0];\mathbb{R}^n),$$

$$M\phi = \int_{-h}^{0} d\mu(\tau)\phi(\tau), \quad \phi \in C([-h,0];\mathbb{R}^m).$$

Corresponding η, β, γ, μ are normalized functions of bounded variation on the interval $[-a,0]$ or $[-h,0]$ with values in $\mathbb{R}^{n\times n}$, $\mathbb{R}^{n\times m}$, $\mathbb{R}^{m\times n}$, $\mathbb{R}^{m\times m}$, respectively.

A function $\alpha : [-T,0] \to \mathbb{R}^k$ of bounded variation on a negative time interval will be called *normalized* if $\alpha(0) = 0$ and if $\alpha(\tau)$ is left continuous for $-T < \tau < 0$. The corresponding function space is denoted by $NBV([-T,0];\mathbb{R}^k)$. Note that, for any $\alpha \in NBV([-T,0];\mathbb{R}^k)$, the function $\tilde{\alpha} : [0,T] \to \mathbb{R}^k$, defined by

$$\tilde{\alpha}(t) = -\alpha(-t), \quad 0 \leq t \leq T,$$

is a normalized function of bounded variation in the sense of Section 1.1.

1.2.1 REMARKS

(i) In the following we extend any function $\alpha : [a,b] \to \mathbb{R}^k$ of bounded variation to the whole real axis by defining $\alpha(t) = \alpha(a)$ for $t \leq a$ and $\alpha(t) = \alpha(b)$ for $t \geq b$.

Any measurable function $x : [a,b] \to \mathbb{R}^k$ will be extended to the whole real axis by defining $x(t) = 0$ for $t \notin [a,b]$.

(ii) At the first glance the right-hand side of (13) seems to be a well-defined expression for $t \geq 0$ only if $w(t)$ and $x(t)$ are continuous (for $t \geq -a$ respectively $t \geq -h$). However, the following equation holds:

$$Lw_t = \int_{-a}^{0} d\eta(\tau)w(t+\tau) = \int_{0}^{a} d\tilde{\eta}(s)w(t-s) = d\tilde{\eta} * w(t)$$

and the last expression makes sense (as an L^p-function on the interval $[0,T]$) for any $w \in L^p([-a,T];\mathbb{R}^n)$. More precisely, in this case the function $t \to Lw_t$ in $L^p([0,T];\mathbb{R}^n)$ is defined by the equation

$$\int_0^T z^T(t)Lw_t dt = \int_{-a}^{0}\int_0^T z^T(t)d\eta(\tau)w(t+\tau)dt$$

for every $z \in L^q([0,T];\mathbb{R}^n)$, $1/p + 1/q = 1$ (compare Remarks 1.1.1 (v) and (xi)).

1.2.2 DEFINITION. *A pair* $w \in L^p([-a,T];\mathbb{R}^n)$, $x \in L^p([-h,T];\mathbb{R}^m)$ *is said to be a solution of* (13) *if* $w(t)$ *is absolutely continuous on* $[0,T]$ *with derivative in* $L^p([0,T];\mathbb{R}^n)$ *and equation* (13) *is satisfied for almost every* $t \in [0,T]$.

We will study the solutions of (13) in the product space

$$M^p = \mathbb{R}^n \times L^p([-a,0];\mathbb{R}^n) \times L^p([-h,0];\mathbb{R}^m)$$

$(1 < p < \infty)$ endowed with the norm

$$\|\phi\| = [|\phi^0|^p + \|\phi^1\|_p^p + \|\phi^2\|_p^p]^{1/p},\ \phi = (\phi^0,\phi^1,\phi^2) \in M^p.$$

This is motivated by the following result.

1.2.3 THEOREM. *Let* η, β, γ, μ *be given as above and suppose that*

$$-1 \notin \sigma(\lim_{\tau\uparrow 0} \mu(\tau)). \tag{14}$$

Moreover let $T > 0$. *Then the following statements hold.*

(i) *For every* $\phi \in M^p$, $f \in L^p([0,T];\mathbb{R}^n)$ *and* $g \in L^p([0,T];\mathbb{R}^m)$, *equation* (13) *admits a unique solution* $w \in L^p([-a,T];\mathbb{R}^n)$, $x \in L^p([-h,T];\mathbb{R}^m)$, *satisfying the initial condition*

$$\begin{aligned} &w(0) = \phi^0,\ w(\tau) = \phi^1(\tau),\ -a \leq \tau < 0, \\ &x(\tau) = \phi^2(\tau),\ -h \leq \tau < 0. \end{aligned} \tag{15}$$

Moreover, the solution operator of (13) *which maps the triple*

$$(\phi,f,g) \in M^p \times L^p([0,T];\mathbb{R}^n) \times L^p([0,T];\mathbb{R}^m)$$

into the pair

$$(w,x) \in W^{1,p}([0,T];\mathbb{R}^n) \times L^p([0,T];\mathbb{R}^m)$$

is bounded and linear.

(ii) *The solution operator of* (13) *which maps the triple* (ϕ,f,g) *into the final state* $(w(T),w_T,x_T) \in M^p$ *is linear, bounded, and compact in* f.

(iii) *If* $T \geqslant a + h$ *and if* $\mu(\tau)$ *is absolutely continuous for* $\tau < 0$, *then this solution operator is compact in* ϕ.

(iv) *Let* $g \in C([0,T];\mathbb{R}^m)$, $\phi^1 \in C([-a,0];\mathbb{R}^n)$, $\phi^2 \in C([-h,0];\mathbb{R}^m)$, $\phi^0 = \phi^1(0)$, *and* $\phi^2(0) = \Gamma\phi^1 + M\phi^2 + g(0)$. *Then the unique solution pair of* (13), (15) *is in* $C([-a,T];\mathbb{R}^n) \times C([-h,T];\mathbb{R}^m)$ *and depends (in this space) continuously on* ϕ, f, *and* g.

(v) *Let* $g \in W^{1,p}([0,T];\mathbb{R}^m)$, $\phi^1 \in W^{1,p}([-a,0];\mathbb{R}^n)$, $\phi^2 \in W^{1,p}([-h,0];\mathbb{R}^m)$, $\phi^0 = \phi^1(0)$, *and* $\phi^2(0) = \Gamma\phi^1 + M\phi^2 + g(0)$. *Then the unique solution pair of* (13), (15) *is in* $W^{1,p}([-a,T];\mathbb{R}^n) \times W^{1,p}([-h,T];\mathbb{R}^m)$ *and depends (in this space) continuously on* ϕ, f, *and* g.

Proof. (ii) We integrate the first equation in (13). For this purpose we need the following identity (compare Remark 1.2.1 (ii)):

$$\int_0^t Lw_s ds = \int_{-a}^0 d\eta(\tau) \int_0^t w(s+\tau)ds = \int_{-a}^0 d\eta(\tau) \int_\tau^{t+\tau} w(s)ds$$

$$= -\eta(-a) \int_{-a}^{t-a} w(s)ds - \int_{-a}^0 \eta(\tau)[w(t+\tau) - w(\tau)]d\tau$$

$$= \int_{-a}^0 \eta(\tau)\phi^1(\tau)d\tau - \int_{-a}^t \eta(s-t)w(s)ds$$

$$= \int_{-a}^0 [\eta(\tau) - \eta(\tau-t)]\phi^1(\tau)d\tau - \int_0^t \eta(s-t)w(s)ds.$$

This expression separates the solution ($t > 0$) of (13) from the initial function ($t \leq 0$). The term $\int_0^t Bx_s ds$ can be transformed by analogy. Hence integration of the first equation in (13) leads to the following equivalent system of Volterra-Stieltjes integral equations

$$
\begin{aligned}
w &= \tilde{\eta} * w + \tilde{\beta} * x + \tilde{f} \\
x &= d\tilde{\gamma} * w + d\tilde{\mu} * x + \tilde{g}
\end{aligned}
\tag{16}
$$

where $w(t)$ and $x(t)$ are now understood as L^p-functions on $[0,T]$. The inhomogeneous terms $\tilde{f} \in C([0,T];\mathbb{R}^n)$ and $\tilde{g} \in L^p([0,T];\mathbb{R}^m)$ are given by

$$
\begin{aligned}
\tilde{f}(t) &= \phi^0 + \int_{-a}^{0} [\eta(\tau) - \eta(\tau-t)]\phi^1(\tau)\, d\tau \\
&\quad + \int_{-h}^{0} [\beta(\tau) - \beta(\tau-t)]\phi^2(\tau)d\tau + \int_0^t f(s)ds, \\
\tilde{g}(t) &= \int_{-a}^{-t} d\gamma(\tau)\phi^1(t+\tau) + \int_{-h}^{-t} d\mu(\tau)\phi^2(t+\tau) + g(t).
\end{aligned}
\tag{17}
$$

It follows from (8) that $\tilde{f}$ and $\tilde{g}$ depend continuously on ϕ and g. Moreover, $\tilde{f}$ depends compactly on f (Remark 1.1.1 (i)). Finally, condition (14) guarantees that equation (16) satisfies the assumptions of Theorem 1.1.4. This proves (ii).

(i) follows from (ii) and from the fact that the right-hand side of the first equation in (13) - as a function in $L^p([0,T];\mathbb{R}^n)$ - depends continuously on $f \in L^p([0,T];\mathbb{R}^n)$, $w \in L^p([-a,T];\mathbb{R}^n)$, and $x \in L^p([-h,T];\mathbb{R}^m)$.

(iii) Let $\mu(\tau)$ be absolutely continuous for $\tau < 0$ and let $g(t) \equiv 0$ and $f(t) \equiv 0$. Moreover, suppose that $\phi^1 \in W^{1,p}([-a,0];\mathbb{R}^n)$ and $\phi^0 = \phi^1(0)$. Then (17) implies

$$
\begin{aligned}
\tilde{g}(t) &= \gamma(-t)\phi^1(0) - \gamma(-a)\phi^1(t-a) - \int_{-a}^{-t} \gamma(\tau)\dot{\phi}^1(t+\tau)d\tau \\
&\quad + \int_{-h}^{-t} d\mu(\tau)\phi^2(t+\tau) \\
&= [\gamma(-t) - \gamma(-a)]\phi^1(0) - \int_{-a}^{-t} [\gamma(\tau) - \gamma(-a)]\dot{\phi}^1(t+\tau)d\tau \\
&\quad + \int_{-h}^{-t} \dot{\mu}(\tau)\phi^2(t+\tau)d\tau, \quad 0 \leq t \leq T.
\end{aligned}
$$

Hence $\tilde{f}$ and $\tilde{g}$ depend compactly on the pair $\phi^1 \in W^{1,p}([-a,0];\mathbb{R}^n)$, $\phi^2 \in L^p([-h,0];\mathbb{R}^m)$. Consequently, the triple $(w(h),w_h,x_h) \in M^p$ also depends compactly on this pair (ϕ^1,ϕ^2). Now it follows from (i) that the composed map

$$M^p \longrightarrow W^{1,p}([-a,0];\mathbb{R}^n) \times L^p([-h,0];\mathbb{R}^m) \longrightarrow M^p$$

$$\phi \longrightarrow (w_a,x_a) \longrightarrow (w(a+h),w_{a+h},x_{a+h})$$

is compact.

(iv) The continuity of $w(t)$ for $t \geqslant -a$ follows from (i) and the fact that $\phi^0 = \phi^1(0)$. Now define

$$\Phi^2(\tau) = \phi^2(\tau),\ -h \leqslant \tau \leqslant 0,\quad \Phi^2(t) = \phi^2(0),\ 0 \leqslant t \leqslant T,$$

$$\tilde{g}(t) = g(t) + \Gamma w_t + M\Phi_t^2 - \phi^2(0),\ 0 \leqslant t \leqslant T,$$

$$\tilde{x}(t) = x(t) - \Phi^2(t),\ -h \leqslant t \leqslant T.$$

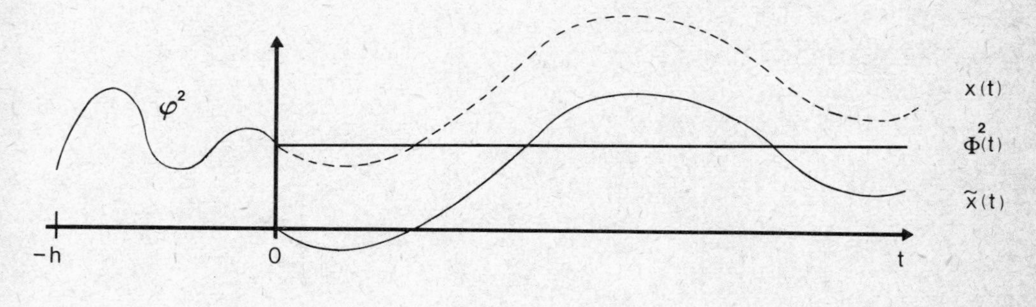

Figure 1

Then it follows from the assumptions of (iv) that $\tilde{g}(t)$ is continuous and $\tilde{g}(0) = 0$. Moreover

$$\tilde{x}(t) = x(t) - \phi^2(0) = \Gamma w_t + Mx_t + g(t) - \phi^2(0)$$

$$= \tilde{g}(t) + d\tilde{\mu} * \tilde{x}(t),\ 0 \leqslant t \leqslant T.$$

Hence, by Theorem 1.1.2 (i), $\tilde{x}(t)$ is continuous for $t \geqslant 0$, satisfies $\tilde{x}(0) = 0$, and depends continuously on $\tilde{g}$. This implies that $x(t)$ is continuous for $t \geqslant -h$.

The proof of (v) is strictly analogous to that of (iv). We obtain that $\tilde{g} \in W^{1,p}([0,T];\mathbb{R}^n)$ and $\tilde{g}(0) = 0$ which allows the application of Corollary 1.1.5. □

The above theorem allows us to define the solution semigroup of the homogeneous equation (13) ($f(t) \equiv 0$ and $g(t) \equiv 0$). Some basic properties of this semigroup are summarized in the next corollary which is a direct consequence of Theorem 1.2.3.

1.2.4 COROLLARY. *Let* μ *satisfy* (14) *and let the operators* $S(t)$ *on* M^p *be defined by*

$$S(t)\phi = (w(t), w_t, x_t), \quad \phi \in M^p,$$

where the pair $w \in L^p_{loc}([-a,\infty);\mathbb{R}^n)$, $x \in L^p_{loc}([-h,\infty);\mathbb{R}^m)$ *is the unique solution of the homogeneous system* (13), (15) ($f(t) \equiv 0$, $g(t) \equiv 0$). *Then the following statements hold.*

(i) $S(t)$ *is a strongly continuous semigroup of bounded linear operators on* M^p.

(ii) *If* $\mu(\tau)$ *is absolutely continuous for* $\tau < 0$, *then* $S(t)$ *is a compact operator for* $t \geqslant a + h$.

(iii) *The restriction of* $S(t)$ *to each of the invariant subspaces*

$$\{\phi \in M^p \,|\, \phi^1 \in C([-a,0];\mathbb{R}^n),\ \phi^2 \in C([-h,0];\mathbb{R}^m),$$
$$\phi^0 = \phi^1(0),\ \phi^2(0) = \Gamma\phi^1 + M\phi^2\},$$

$$\{\phi \in M^p \,|\, \phi^1 \in W^{1,p}([-a,0];\mathbb{R}^n),\ \phi^2 \in W^{1,p}([-h,0];\mathbb{R}^m),$$
$$\phi^0 = \phi^1(0),\ \phi^2(0) = \Gamma\phi^1 + M\phi^2\}$$

is a C_0*-semigroup in the respective topology.*

The solution of the inhomogeneous equation (13) with $f(t) \not\equiv 0$ and $g(t) \equiv 0$ can be described by a variation-of-constants formula in the Banach space M^p. For the proof of this result we need the fact that the dual space of M^p can be identified with M^q via the pairing

$$\langle\psi,\phi\rangle = \psi^{0^T}\phi^0 + \int_{-a}^{0} \psi^{1^T}(\tau)\phi^1(\tau)d\tau + \int_{-h}^{0} \psi^{2^T}(\tau)\phi^2(\tau)d\tau$$

for $\phi \in M^p$ and $\psi \in M^q$.

1.2.5 THEOREM. *Let* $w \in L^p([-a,T];\mathbb{R}^n)$ *and* $x \in L^p([-h,T];\mathbb{R}^m)$ *be the unique solution of* (13), (15) *corresponding to* $\phi = 0$, $g \equiv 0$, *and* $f \in L^p([0,T];\mathbb{R}^m)$. *Then, for every* $t \in [0,T]$,

$$(w(t),w_t,x_t) = \int_0^t S(t-s)(f(s),0,0)ds \in M^p.$$

Proof. Let $\rho \in NBV([0,T);\mathbb{R}^{n\times n})$ be defined by $\rho(0) = 0$ and $\rho(t) = I$ for $0 < t < T$. Moreover, define $W \in NBV[0,T);\mathbb{R}^{n\times n})$ and $X \in NBV([0,T);\mathbb{R}^{m\times n})$ to be the unique solution of the Volterra-Stieltjes integral equation

$$W = \rho + \tilde{\eta} * W + \tilde{\beta} * X,$$

$$X = d\tilde{\gamma} * W + d\tilde{\mu} * X$$

(Theorem 1.1.2 (ii)). Then $W(t)$ and $X(t)$ form the left n columns of the fundamental solution of (16) in the sense of Definition 1.1.3.

Now let $w(t;\phi^0,f)$, $-a \leqslant t \leqslant T$, and $x(t;\phi^0,f)$, $-h \leqslant t \leqslant T$, be the unique solution of (13), (15) with $\phi^1 = 0$, $\phi^2 = 0$, and $g \equiv 0$. Then it follows from Theorem 1.1.4 - applied to (16) and (17) - that the equations

$$w(t;\phi^0,f) = \int_0^t dW(s)[\phi^0 + \int_0^{t-s} f(\tau)d\tau]$$

$$x(t;\phi^0,f) = \int_0^t dX(s)[\phi^0 + \int_0^{t-s} f(\tau)d\tau]$$

hold for almost every $t \in [0,T]$ (note that these equations are trivially satisfied for $t \leqslant 0$). For $f \equiv 0$ this implies $w(t;\phi^0,0) = W(t)\phi^0$ and $x(t;\phi^0,0) = X(t)\phi^0$, which means that

$$S(t)(\phi^0,0,0) = (W(t),W_t,X_t)\phi^0, \quad 0 < t < T.$$

On the other hand we obtain in the case $\phi^0 = 0$

$$w(t) = w(t;0,f) = \int_0^t W(t-s)f(s)ds, \quad -a \leqslant t < T,$$

$$x(t) = x(t;0,f) = \int_0^t X(t-s)f(s)ds, \quad -h \leqslant t < T.$$

Hence the following equation holds for every $\psi \in M^q$ and $t \geqslant 0$

$$\langle\psi,(w(t),w_t,x_t)\rangle$$

$$= \psi^{0^T} \int_0^t W(t-s)f(s)ds + \int_{-a}^0 \psi^{1^T}(\tau) \int_0^{t+\tau} W(t+\tau-s)f(s)ds\, d\tau$$

$$+ \int_{-h}^0 \psi^{2^T}(\tau) \int_0^{t+\tau} X(t+\tau-s)f(s)ds\, d\tau$$

$$= \int_0^t \psi^{0^T} W(t-s)f(s)ds + \int_0^t \int_{-a}^0 \psi^{1^T}(\tau)W(t-s+\tau)f(s)d\tau\, ds$$

$$+ \int_0^t \int_{-h}^0 \psi^{2^T}(\tau)X(t-s+\tau)f(s)d\tau\, ds$$

$$= \int_0^t \langle\psi,\mathcal{S}(t-s)(f(s),0,0)\rangle\, ds$$

$$= \langle\psi, \int_0^t \mathcal{S}(t-s)(f(s),0,0)ds\rangle. \qquad \square$$

We close this section with some results on the infinitesimal generator of $\mathcal{S}(t)$.

1.2.6 THEOREM. *The infinitesimal generator of* $\mathcal{S}(t)$ *is given by*

$$\text{dom } A = \{\phi \in M^p | \phi^1 \in W^{1,p}([-a,0];\mathbb{R}^n),\ \phi^2 \in W^{1,p}([-h,0];\mathbb{R}^m),$$

$$\phi^0 = \phi^1(0),\ \phi^2(0) = \Gamma\phi^1 + M\phi^2\},$$

$$A\phi = (L\phi^1 + B\phi^2,\ \dot{\phi}^1,\ \dot{\phi}^2) \in M^p.$$

Proof. Let A be the operator defined above. Moreover, for any $\phi \in M^p$, let $w(t;\phi)$, $t \geqslant -a$, and $x(t;\phi)$, $t \geqslant -h$, denote the corresponding solution of the homogeneous system (13), (15).

First let $\phi \in \text{dom } A$. Then, by Theorem 1.2.3 (v), $w(\cdot;\phi) \in W^{1,p}([-a,T];\mathbb{R}^n)$ and $x(\cdot;\phi) \in W^{1,p}([-h,T];\mathbb{R}^m)$ for every $T \geqslant 0$. This implies that $w(t;\phi)$ is continuously differentiable for $t \geqslant 0$ and $\dot{w}(0,\phi) = L\phi^1 + B\phi^2$. Hence the existence of the limit in M^p, $\lim_{t\downarrow 0} t^{-1}[S(t)\phi - \phi] = A\phi$, follows from

$$\int_{-a}^{0} \left|\frac{w(t+\tau;\phi) - \phi^1(\tau)}{t} - \dot{\phi}^1(\tau)\right|^p d\tau \leqslant \sup_{0\leqslant s\leqslant t} \int_{-a}^{0} |\dot{w}(s+\tau;\phi) - \dot{\phi}^1(\tau)|^p d\tau$$

and an analogous inequality for ϕ^2 (compare Bernier-Manitius [11, Appendix]).

Conversely, let $\phi \in M^p$ be in the domain of the infinitesimal generator of $S(t)$ and define

$$\Phi = \lim_{t\downarrow 0} t^{-1}[S(t)\phi - \phi] \in M^p.$$

Then the following equation holds for almost every $\tau \in [-a,0]$:

$$\begin{aligned}
\phi^0 - \phi^1(\tau) &= \lim_{t\downarrow 0} t^{-1}\left(\int_0^t w(s;\phi)ds - \int_\tau^{t+\tau} w(s;\phi)ds\right)\\
&= \lim_{t\downarrow 0} t^{-1}\left(\int_{t+\tau}^t w(s;\phi)ds - \int_\tau^0 w(s;\phi)ds\right)\\
&= \lim_{t\downarrow 0} \int_\tau^0 t^{-1}\Big(w(t+\sigma;\phi) - \phi^1(\sigma)\Big)d\sigma\\
&= \int_\tau^0 \Phi^1(\sigma)d\sigma.
\end{aligned}$$

Hence $\phi^1 \in W^{1,p}([-a,0];\mathbb{R}^n)$, $\phi^1(0) = \phi^0$, and $\dot{\phi}^1 = \Phi^1$. By analogy, we obtain for almost all τ, $\theta \in [-h,0)$

$$\begin{aligned}
\phi^2(\tau) - \phi^2(\theta) &= \lim_{t\downarrow 0} t^{-1}\left(\int_\tau^{t+\tau} x(s;\phi)ds - \int_\theta^{t+\theta} x(s;\phi)ds\right)\\
&= \lim_{t\downarrow 0} \int_\theta^\tau t^{-1}\Big(x(t+\sigma;\phi) - \phi^2(\sigma)\Big)d\sigma\\
&= \int_\theta^\tau \Phi^2(\sigma)d\sigma.
\end{aligned}$$

This shows that $\phi^2 \in W^{1,p}([-h,0];\mathbb{R}^m)$ and $\dot{\phi}^2 = \Phi^2$. Now it is known from general semigroup theory that $S(t)\phi$ is in the domain of the generator for every $t \geqslant 0$. In particular, the function $\tau \to x(t+\tau;\phi)$ is continuous on $[-h,0]$. Taking $t < h$ and $\tau = -t$, we obtain

$$\phi^2(0) = x(0;\phi) = \Gamma\phi^1 + M\phi^2.$$

Hence $\phi \in \text{dom } A$. □

In the following we replace the state space M^p and the operator A by their obvious complex extensions. Then the spectrum of A can be characterized by the complex $(n+m) \times (n+m)$-matrix function

$$\Delta(\lambda) = \begin{bmatrix} \lambda I - L(e^{\lambda\cdot}) & -B(e^{\lambda\cdot}) \\ -\Gamma(e^{\lambda\cdot}) & I-M(e^{\lambda\cdot}) \end{bmatrix}, \ \lambda \in \mathbb{C}, \tag{18}$$

where $e^{\lambda\cdot}$ denotes the function $\tau \to e^{\lambda\tau}$ on the interval $[-a,0]$ respectively $[-h,0]$. For the proof of this result we need the convolution of two functions g and f on a negative time-interval $[-T,0]$, given by

$$g * f(\tau) = \int_\tau^0 g(\tau-\sigma)f(\sigma)d\sigma, \ -T \leqslant \tau \leqslant 0.$$

1.2.7 THEOREM. *Let* $\lambda \in \mathbb{C}$, ϕ, $\Phi \in M^p$, *and let* $\Delta(\lambda)$ *be given by* (18). *Then the following statements hold.*

(i) $\phi \in \text{dom } A$ *and* $(\lambda I - A)\phi = \Phi$ *if and only if*

$$\begin{aligned} \phi^1(\tau) &= e^{\lambda\tau}\phi^0 + \int_\tau^0 e^{\lambda(\tau-\sigma)}\Phi^1(\sigma)d\sigma, \ -a \leqslant \tau \leqslant 0, \\ \phi^2(\tau) &= e^{\lambda\tau}\phi^2(0) + \int_\tau^0 e^{\lambda(\tau-\sigma)}\Phi^2(\sigma)d\sigma, \ -h \leqslant \tau \leqslant 0, \end{aligned} \tag{19.1}$$

and

$$\Delta(\lambda)\begin{pmatrix} \phi^0 \\ \phi^2(0) \end{pmatrix} = \begin{pmatrix} \Phi^0 + L(e^{\lambda\cdot} * \Phi^1) + B(e^{\lambda\cdot} * \Phi^2) \\ \Gamma(e^{\lambda\cdot} * \Phi^1) + M(e^{\lambda\cdot} * \Phi^2) \end{pmatrix}. \tag{19.2}$$

(ii) $\lambda \in \sigma(A) = P\sigma(A)$ *if and only if* $\det \Delta(\lambda) = 0$.

(iii) *If* $\lambda \notin \sigma(A)$, *then the resolvent operator* $(\lambda I - A)^{-1}$ *of* A *on* M^p *is compact.*

Proof. (i) It follows from Theorem 1.2.6 that $\phi \in \text{dom } A$ and $(\lambda I - A)\phi = \Phi$ if and only if ϕ^1 and ϕ^2 are absolutely continuous, $\phi^0 = \phi^1(0)$, and

$$\phi^2(0) - \Gamma\phi^1 - M\phi^2 = 0,$$

$$\lambda\phi^0 - L\phi^1 - B\phi^2 = \Phi^0,$$

$$\dot{\phi}^1(\tau) = \lambda\phi^1(\tau) - \Phi^1(\tau), \ -a \leqslant \tau \leqslant 0,$$

$$\dot{\phi}^2(\tau) = \lambda\phi^2(\tau) - \Phi^2(\tau), \ -h \leqslant \tau \leqslant 0.$$

The last two equations are equivalent to (19.1). If (19.1) is satisfied, then the first two of these equations are equivalent to (19.2).

(ii) In the case $\Phi = 0$ statement (i) shows that $\lambda I - A$ is injective (i.e. $\lambda \notin P\sigma(A)$) if and only if $\det \Delta(\lambda) \neq 0$. Moreover, if $\det \Delta(\lambda) \neq 0$, then it follows again from statement (i) that $\lambda I - A$ is bijective (i.e. $\lambda \notin \sigma(A)$).

(iii) Compactness of the resolvent operator follows from Statement (i) and Remark 1.1.1 (ii). □

1.3 UNBOUNDED CONTROL AND OBSERVATION FOR INFINITE-DIMENSIONAL LINEAR SYSTEMS

In this section we treat the abstract evolution equation

$$d/dt\, x(t) = Ax(t) + Bu(t) \tag{20}$$

in two reflexive Banach spaces X and X where X is embedded into X as a dense subspace. The desired state space of system (20) will be X. However, the input operator B acts in the bigger space X.

Correspondingly, we want to study the observed system

$$d/dt\, x(t) = Ax(t), \ y(t) = Cx(t), \tag{21}$$

in the state space X while the output operator C is only defined on the subspace X.

In particular, we prove perturbation results which arise in state feedback for system (20) respectively in output injection for system (21). Some of these results can also be proved for nonreflexive Banach spaces. However, it will be enough for our purposes to study the reflexive case which at some places simplifies the statements and proofs.

We need the following assumptions on the operator A and the Banach spaces X and X.

(H1) Let X be a real reflexive Banach space and A the infinite infinitesimal generator of a strongly continuous semigroup $S(t) : X \to X$. Moreover, we assume that $\mathcal{X}$ is a real reflexive Banach space and $\iota: \mathcal{X} \to X$ an embedding (i.e. a bounded, linear, one-to-one mapping) such that $\operatorname{ran} \iota = \operatorname{dom} A$.

The second part of the above hypothesis means that the Banach space $\mathcal{X}$ is nothing else than the domain of the generator A, endowed with a certain norm. We will see that this norm is equivalent to the graph norm of A.

1.3.1 REMARKS

(i) If necessary, all spaces and operators will in the following be interpreted as their obvious complex extensions.

(ii) Let $\lambda \in \mathbb{C}$. Then it follows from (H1) and the closed graph theorem that the composed operator

$$T_\lambda := (\lambda I - A)\iota : \mathcal{X} \to X$$

is bounded. If moreover $\lambda \notin \sigma(A)$, then this operator is an isomorphism.

(iii) It follows from (ii) that the usual norm $\|\cdot\|_{\mathcal{X}}$ on $\mathcal{X}$ is equivalent to the graph norm

$$\|x\|_A = \|\iota x\|_X + \|A\iota x\|_X, \; x \in \mathcal{X},$$

of A. In fact, we have

$$\|x\|_A \leqslant [\|\iota\|_{L(\mathcal{X},X)} + \|A\iota\|_{L(\mathcal{X},X)}] \; \|x\|_{\mathcal{X}}$$

and for every $\lambda \notin \sigma(A)$

$$\|x\|_{\mathcal{X}} \leqslant \|[(\lambda I - A)\iota]^{-1}\|_{L(X,\mathcal{X})} \; \|(\lambda I - A)\iota x\|_X$$

$$\leqslant \|[(\lambda I - A)\iota]^{-1}\|_{L(X,\mathcal{X})} \max\{|\lambda|,1\} \; \|x\|_A .$$

(iv) Since the Banach spaces X and $\mathcal{X}$ are reflexive, the mapping $\iota^*: X^* \to \mathcal{X}^*$ is an embedding of X^* into $\mathcal{X}^*$ as a dense subspace.

(v) Let $x^* \in \mathcal{X}^*$ be given. Then there exists an $x^* \in X^*$ such that $\|x^*\| \leqslant K$ and $\iota^* x^* = x^*$ if and only if the inequality

$$\langle x^*, x\rangle \leqslant K\|\iota x\|_X$$

holds for all $\in X$ ($x^* \in X^*$ can be constructed via continuous extension of the map $\iota x \to \langle x^*, x\rangle$).

(vi) Let $x \in X$ be given. Then there exists an $x \in X$ such that $\|x\|_X \leqslant K$ and $\iota x = x$ if and only if the inequality

$$\langle x^*, x\rangle \leqslant K\,\|\iota^* x^*\|_{X^*}$$

holds for all $x^* \in X^*$.

It follows from the commutativity of $S(t)$ and A that the restriction of $S(t)$ to dom A is a strongly continuous family of bounded linear operators with respect to the graph norm. The semigroup property is obviously satisfied. Hence there exists a unique C_0-semigroup, $S(t) : X \to X$ such that the following diagram commutes

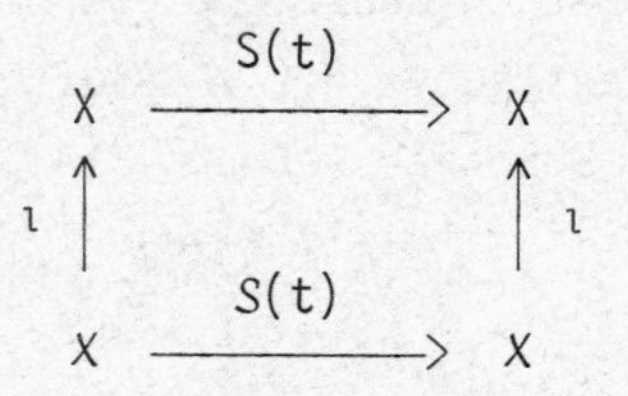

This means that

$$\iota S(t) = S(t)\iota. \tag{22}$$

The infinitesimal generator of this semigroup has the following properties.

1.3.2 <u>LEMMA</u>. *Let* (H1) *be satisfied and* $S(t)$ *defined as above. Then the following statements hold.*

(i) *The infinitesimal generator of* $S(t)$ *is given by*

$$\iota A x = A\iota x,\ \text{dom } A = \{x \in X \mid A\iota x \in \text{dom } A = \text{ran } \iota\}. \tag{23}$$

(ii) $P\sigma(A) = P\sigma(A)$, $\sigma(A) = \sigma(A)$.

(iii) *Let* $\mu \notin \sigma(A)$ *and define* $T_\mu = (\mu I - A)\iota : X \to X$. *Then*

$$S(t) = T_\mu S(t) T_\mu^{-1},\ t \geqslant 0.$$

(iv) dom $A^* = \operatorname{ran} \iota^*$.

(v) $A^*\iota^* = (A\iota)^* \in L(X^*,X^*)$

Proof. (i) Let the operator A be defined by (23). Moreover let x, $w \in X$ be given. Then it follows from Remark 1.3.1 (iii) that the limit $w = \lim_{t\downarrow 0} t^{-1}[S(t)x - x]$ exists in X if and only if

$$\lim_{t\downarrow 0} \left\|\frac{S(t)\iota x - \iota x}{t} - \iota w\right\|_X = \lim_{t\downarrow 0} \left\|\frac{S(t)A\iota x - A\iota x}{t} - A\iota w\right\|_X = 0.$$

But this is equivalent to $A\iota x \in \operatorname{dom} A = \operatorname{ran} \iota$ and $A\iota x = \iota w$. By (23), this means that $x \in \operatorname{dom} A$ and $Ax = w$.

(ii) It follows from (i) that $x \in \ker(\lambda I - A)$ if and only if $\iota x \in \ker(\lambda I - A)$. Hence $P\sigma(A) = P\sigma(A)$.

Now let $\lambda I - A$ be onto and let $x \in X$. Then there exists some $w \in X$ such that $\iota x = (\lambda I - A)\iota w$. This implies that $A\iota w \in \operatorname{dom} A$ and hence $w \in \operatorname{dom} A$. Moreover $\iota(\lambda I - A)w = (\lambda I - A)\iota w = \iota x$ which proves that $x = (\lambda I - A)w$.

Conversely, let $\lambda I - A$ be onto and let $x \in X$. Moreover let $\mu \notin \sigma(A)$. Then there exists an $x \in \operatorname{dom} A$ such that $(\lambda I - A)x = T_\mu^{-1}x$, this implies

$$x = T_\mu(\lambda I - A)x = (\mu I - A)\iota(\lambda I - A)x = (\lambda I - A)(\mu I - A)\iota x.$$

We conclude that $\lambda I - A$ is onto if and only if $\lambda I - A$ is onto and thus $\sigma(A) = \sigma(A)$.

(iii) Let $\mu \in \mathbb{C}$. Then the following equation holds for every $x \in X$ and $t \geqslant 0$

$$S(t)T_\mu x = (\mu I - A)S(t)\iota x = (\mu I - A)\iota S(t)x = T_\mu S(t)x.$$

(iv), (v) Let $x^* \in X^*$ be given. Then the following equation holds for every $x \in \operatorname{dom} A$

$$\langle \iota^* x^*, Ax\rangle = \langle x^*, \iota Ax\rangle = \langle x^*, A\iota x\rangle = \langle (A\iota)^* x^*, x\rangle.$$

This implies that $\iota^* x^* \in \operatorname{dom} A^*$ and $A^*\iota^* x^* = (A\iota)^* x^*$.

Conversely, let $x^* \in \operatorname{dom} A^*$ and $\mu \notin \sigma(A)$. Then, for every $x \in X$, we have

$$w_x = T_\mu^{-1}\iota x \in \operatorname{dom} A,$$

since $A\iota w_x = \mu\iota w_x - \iota x \in \operatorname{dom} A$. This implies that $(\mu I - A)w_x = x$ and hence

$$\langle x^*,x\rangle = \langle x^*,(\mu I - A)w_x\rangle = \langle(\bar{\mu}I - A^*)x^*,T_\mu^{-1}\iota x\rangle$$

$$\leqslant \|(\bar{\mu}I - A^*)x^*\|_{\mathcal{X}^*} \; \|T_\mu^{-1}\|_{L(X,X)} \; \|\iota x\|_X.$$

We conclude that $x^* \in \operatorname{ran} \iota^*$ (Remark 1.3.1 (v)). □

It follows from Remark 1.3.1 and Lemma 1.3.2 that, if A satisfies (H1), then A^* also satisfies (H1) with X replaced by $\mathcal{X}^*$ and $\mathcal{X}$ by X^*. This situation may be illustrated by the following two commuting diagrams where the diagram on the right-hand side is both the dual and the analogue of the diagram on the left.

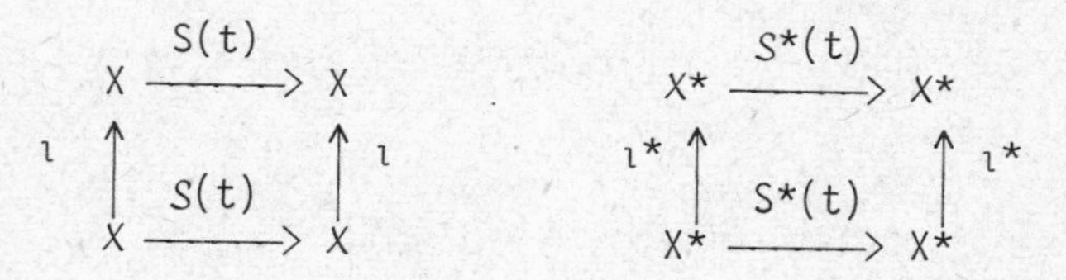

Now we are going to study the solutions of the Cauchy problem (20) where U is a real reflexive Banach space and $B \in L(U,X)$. To do this we need some integration in Banach spaces in the sense of Bochner (see, e.g., Hille-Phillips [53, Sections 3.7 and 3.8] and Dinculeanu [34]). We will make use of some basic properties of the Bochner integral without giving each time an explicit reference. However, we recall that in the applications the input space U is always finite-dimensional, which at some places simplifies the interpretation.

Let us first make precise what we mean by a solution of the Cauchy problem (20) in the state space $\mathcal{X}$.

1.3.3 DEFINITION. *Let* $u(\cdot) \in L^p([0,T];U)$ *be given. Then a continuous function* $x : [0,T] \to \mathcal{X}$ *is said to be a solution of* (20) *if the function* $x(t) = \iota x(t) \in X$ *is absolutely continuous on* $[0,T]$ *and satisfies* (20). *This means that*

$$\iota x(t) = \iota x(0) + \int_0^t [A\iota x(s) + Bu(s)]ds, \quad 0 \leqslant t \leqslant T. \tag{24}$$

The following assumption on the operator B is needed in order to obtain solutions of (20) in the desired state space $\mathcal{X}$.

(H2) Let U be a real reflexive Banach space, $B \in L(U,X)$ and $1 \leq p < \infty$. Moreover, suppose that, for every $T > 0$, there exists some $b_T > 0$ such that

$$\int_0^T S(t)Bu(t)dt \in \operatorname{ran} \iota$$

and

$$\| \iota^{-1} \int_0^T S(t)Bu(t)dt \|_X \leq b_T \|u\|_{p,T} \tag{25}$$

for every $u \in L^p([0,T];U)$ where

$$\|u\|_{p,T} = [\int_0^T \|u(t)\|_U^p \, dt]^{1/p}.$$

1.3.4 THEOREM. *Let* (H1) *and* (H2) *be satisfied and let* $x_0 \in X$ *be given. Then*

$$x(t) = S(t)x_0 + \iota^{-1} \int_0^t S(t-s)Bu(s)ds, \; 0 \leq t \leq T, \tag{26}$$

is the unique solution of (20) *(in the sense of Definition 1.3.3) with* $x(0) = x_0$.

Proof. The uniqueness part follows from the fact that the difference $x(t)$ of two solutions of (20) corresponding to the same input $u(\cdot)$ and the same initial state x_0 satisfies

$$\iota x(t) = \int_0^t A\iota x(s)ds, \; 0 \leq t \leq T.$$

Since $A\iota : X \to X$ is a bounded operator (Remark 1.3.1 (ii)), we obtain that $x(t) = \iota x(t) \in X$ is continuously differentiable on $[0,T]$ and satisfies $\dot{x}(t) = Ax(t)$, $x(0) = 0$. Hence it follows from classical results in semigroup theory (see, e.g., Goldstein [39], Pazy [127]) that $x(t) = 0$ and thus $x(t) = 0$ for $0 \leq t \leq T$.

Now let $x(t)$ be given by (26). Moreover let $t, s \in [0,T]$ and define $u(\tau) := 0$ for $\tau \notin [0,T]$. Then, by (25), we have

$$\|x(t) - x(s)\|_X$$

$$\leq \|S(t)x_0 - S(s)x_0\|_X + \|\iota^{-1}\int_0^T S(\tau)B[u(t-\tau) - u(s-\tau)]d\tau\|_X$$

$$\leq \|S(t)x_0 - S(s)x_0\|_X + b_T[\int_0^T \|u(t-\tau) - u(s-\tau)\|_U^p \, d\tau]^{1/p}$$

and hence $x(t)$ is continuous. Moreover, it is well known that, for every continuously differentiable input $u(t)$, the function

$$x(t) = \iota x(t) = S(t)\iota x_0 + \int_0^t S(t-s)Bu(s)ds, \; 0 \leq t \leq T,$$

is continuously differentiable and satisfies (20) (see, e.g., Curtain-Pritchard [24], Goldstein [39], Pazy [127]). Hence (24) is satisfied in this case. In general, (24) follows from the fact that both sides of this equation depend continuously on $u(\cdot) \in L^p([0,T];U)$. □

Now we come to the assumption on the output operator C which guarantees the existence of an output function of system (21) for every initial state in X.

(H3) Let Y be a real reflexive Banach space, $C \in L(X,Y)$, and $1 < q \leq \infty$. Moreover, suppose that, for every $T > 0$, there exists some $c_T > 0$ such that the following inequality holds for every $x \in X$

$$\|CS(\cdot)x\|_{q,T} \leq c_T \; \|\iota x\|_X. \tag{27}$$

This hypothesis is actually the dual of (H2) as is shown in the lemma below.

1.3.5 LEMMA. *Let* (H1) *be satisfied and let* U *be a real reflexive Banach space. Then* $B \in L(U,X)$ *satisfies* (H2) *if and only if the inequality*

$$\|B^*S^*(\cdot)x^*\|_{q,T} \leq b_T \; \|\iota^*x^*\|_{X^*} \tag{28}$$

holds for all $x^* \in X^*$ *and* $T > 0$ $(1/p + 1/q = 1)$.

Proof. The following equation holds for all $x^* \in X^*$ and all $u(\cdot) \in L^p([0,T];U)$

$$\langle B^*S^*(\cdot)x^*, u(\cdot)\rangle_{L^q([0,T];U^*),L^p([0,T];U)}$$

$$= \int_0^T \langle B^*S^*(t)x^*, u(t)\rangle_{U^*,U}\, dt$$

$$= \langle x^*, \int_0^T S(t)Bu(t)dt\rangle_{X^*,X}.$$

Hence (H2) implies that, given $x^* \in X^*$, the inequality

$$\langle B^*S^*(\cdot)x^*, u(\cdot)\rangle_{L^q,L^p} = \langle \iota^*x^*, \iota^{-1}\int_0^T S(t)Bu(t)dt\rangle_{X^*,X}$$

$$\leq b_T\, \|\iota^*x^*\|_X\, \|u(\cdot)\|_{p,T}$$

holds for all $u(\cdot) \in L^p([0,T];U)$. We conclude that (28) is satisfied (Dinculeanu [34, Proposition 14.29]).

Conversely, (28) implies that, given $u(\cdot) \in L^p([0,T];U)$, the inequality

$$\langle x^*, \int_0^T S(t)Bu(t)dt\rangle_{X^*,X} \leq b_T\, \|\iota^*x^*\|_{X^*}\, \|u(\cdot)\|_{p,T}$$

holds for all $x^* \in X^*$. Hence it follows from Remark 1.3.1 (vi) that (H2) is satisfied. □

The next lemma is the basic tool for our perturbation result.

1.3.6 LEMMA. *Let* (H1), (H2) *be satisfied and* $F \in L(X,U)$. *Then, for every* $w(\cdot) \in C([0,T];X)$, *there exists a unique solution* $x(\cdot) \in C([0,T];X)$ *of*

$$x(t) = w(t) + \iota^{-1}\int_0^t S(t-s)BFx(s)ds,\ 0 \leq t \leq T,$$

depending continuously on $w(\cdot)$.

Proof. By Theorem 1.3.4, the expression

$$[Lx](t) = \iota^{-1}\int_0^t S(t-s)BFx(s)ds,\ 0 \leq t \leq T,$$

defines a bounded linear operator on $C([0,T];X)$.

Now choose $\varepsilon > 0$ and $\gamma > 0$ such that

$$\left(\varepsilon^{1/p} + T^{1/p} e^{-\gamma\varepsilon}\right) b_T \;\|F\|_{L(X,U)} < 1$$

and introduce on $C([0,T];X)$ the equivalent norm

$$\|x(\cdot)\|_\gamma = \sup_{0 \leq t \leq T} \|x(t)\|_X \, e^{-\gamma t}.$$

Then, for every $x(\cdot) \in C([0,T];X)$ and every $t \in [0,T]$, the following inequality holds

$$\|\iota^{-1} \int_0^t S(s)BFx(s)ds\|_X \leq b_T \|F\| \left(\int_0^t \|x(s)\|_X^p \, ds\right)^{1/p}$$

$$\leq b_T \|F\| \; t^{1/p} \sup_{0 \leq s \leq t} \|x(s)\|_X .$$

This implies

$$\|[Lx](t)\|_X \, e^{-\gamma t}$$

$$\leq \|\iota^{-1} \int_0^\varepsilon S(s)BFx(t-s)ds\|_X \, e^{-\gamma t}$$

$$+ \|\iota^{-1} \int_\varepsilon^t S(s)BFx(t-s)ds\|_X \, e^{-\gamma t}$$

$$\leq \varepsilon^{1/p} b_T \;\|F\| \sup_{0 \leq s \leq \varepsilon} \|x(t-s)\|_X \, e^{-\gamma t}$$

$$+ t^{1/p} b_T \;\|F\| \sup_{\varepsilon \leq s \leq t} \|x(t-s)\|_X \, e^{-\gamma(t-\varepsilon)} \, e^{-\gamma\varepsilon}$$

$$\leq \left(\varepsilon^{1/p} + T^{1/p} e^{-\gamma\varepsilon}\right) b_T \;\|F\|_{L(X,U)} \;\|x(\cdot)\|_\gamma .$$

We conclude that L is a contraction with respect to $\|\cdot\|_\gamma$ and hence $I - L$ is boundedly invertible. □

Now we are in a position to prove the desired perturbation results. Theorem 1.3.7 is related to the state feedback problem (hypothesis (H2)) and the dual result, Theorem 1.3.9, to the output injection problem (hypothesis (H3)).

1.3.7 <u>THEOREM</u>. *Let* (H1), (H2) *be satisfied and let* $F \in L(X,U)$. *Then the following statements hold.*

(i) *There exists a unique* C_0*-semigroup* $S_F(t) : X \to X$ *such that the equation*

$$S_F(t)x = S(t)x + \iota^{-1} \int_0^t S(t-s)BFS_F(s)x\,ds \tag{29}$$

holds for every $x \in X$ *and every* $t \geq 0$.

(ii) *For every* $x \in X$ *the function* $t \to \iota S_F(t)x$ *is continuously differentiable in* X *and satisfies*

$$d/dt\ \iota S_F(t)x = [A\iota + BF]S_F(t)x,\ t \geq 0.$$

(iii) *The infinitesimal generator of* $S_F(t)$ *is given by*

$$\begin{aligned} &\text{dom } A_F = \{x \in X | A\iota x + BFx \in \text{ran } \iota\}, \\ &\iota A_F x = A\iota x + BFx. \end{aligned} \tag{30}$$

This means that the following diagram commutes

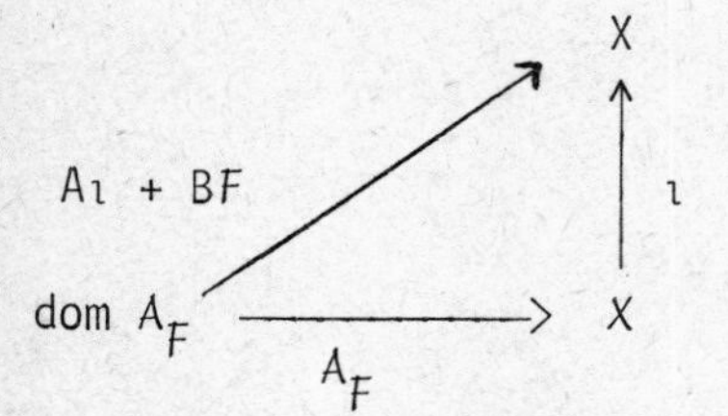

<u>Proof</u>. (i) For every $x_0 \in X$ let $x(t;x_0)$, $t \geq 0$, be the unique solution of

$$x(t;x_0) = S(t)x_0 + \iota^{-1} \int_0^t S(t-s)BFx(s;x_0)ds.$$

Then the operators $S_F(t) : X \to X$, defined by $S_F(t)x_0 = x(t;x_0)$ for $t \geq 0$ and $x_0 \in X$, are bounded and strongly continuous (Lemma 1.3.6) and satisfy (29).

Moreover, the following equation holds for all $x_0 \in X$ and $t, s \geq 0$

$$x(t+s;x_0) = S(t)S(s)x_0 + \iota^{-1}\int_0^{t+s} S(t+s-\tau)BFx(\tau;x_0)d\tau$$

$$= S(t)S(s)x_0 + \iota^{-1} S(t)\int_0^{s} S(s-\tau)BFx(\tau;x_0)d\tau$$

$$+ \iota^{-1}\int_s^{t+s} S(t+s-\tau)BFx(\tau;x_0)d\tau$$

$$= S(t)x(s;x_0) + \iota^{-1}\int_0^{t} S(t-\tau)BFx(\tau+s;x_0)d\tau.$$

Hence, by Lemma 1.3.6, we have $x(t+s;x_0) = x(t;x(s;x_0))$ which proves the semigroup property.

(ii) follows directly from (29) and Theorem 1.3.4.

(iii) Let A_F be the infinitesimal generator of $S_F(t)$. Then the following equation holds for every $x \in \text{dom } A_F$

$$A\iota x + BFx = \lim_{t\downarrow 0} t^{-1}[\iota S_F(t)x - \iota x] = \iota A_F x.$$

Conversely, let $x \in X$ be given such that $A\iota x + BFx \in \text{ran } \iota$ and choose $w \in X$ such that $\iota w = A\iota x + BFx$. Then the function

$$x(t) = x + \int_0^t S_F(s)w ds \in X,\ t \geq 0,$$

is continuous and satisfies the equation

$$\iota x(t) = \iota x + \iota\int_0^t S(s)w ds + \iota\int_0^t \iota^{-1}\int_0^s S(\tau)BFS_F(s-\tau)w d\tau ds$$

$$= \iota x + \int_0^t S(s)\iota w ds + \int_0^t S(\tau)BF\int_\tau^t S_F(s-\tau)w ds d\tau$$

$$= \iota x + \int_0^t S(s)[A\iota x + BFx]ds + \int_0^t S(\tau)BF\int_0^{t-\tau} S_F(s)w ds d\tau$$

$$= S(t)\iota x + \int_0^t S(\tau)BF[x + \int_0^{t-\tau} S_F(s)w ds]d\tau$$

$$= \iota S(t)x + \int_0^t S(t-s)BFx(s)ds.$$

Hence, by Lemma 1.3.6, we have $x(t) = S_F(t)x$ and thus

$$\lim_{t \downarrow 0} \|t^{-1}[S_F(t)x - x] - w\|_X = 0.$$

This means that $x \in \text{dom } A_F$ and $A_F x = w$. □

In order to prove the dual result, or more precisely the relation between both results (Statement (iv) in Theorem 1.3.9), we need the following general semigroup theoretic fact. In the case $X_1 = X_2$ this has been shown by Bernier and Manitius [11, Lemma 5.3]. We present a simplified proof.

1.3.8 LEMMA. *Let* X_1, X_2 *be Banach-spaces and* $S_1(t)$, $S_2(t)$ C_o*-semigroups on* X_1, X_2. *Moreover let* $T \in L(X_1,X_2)$. *Then the following statements are equivalent.*

(i) $S_2(t)T = TS_1(t)$, $t \geqslant 0$.

(ii) *For every* $x \in \text{dom } A_1$ *we have* $Tx \in \text{dom } A_2$ *and* $A_2Tx = TA_1x$.

Proof. First let (i) be satisfied and $x \in \text{dom } A_1$. Then

$$TA_1x = \lim_{t \downarrow 0} \frac{TS_1(t)x - Tx}{t} = \lim_{t \downarrow 0} \frac{S_2(t)Tx - Tx}{t}$$

which proves (ii).

Conversely, let (ii) be satisfied and $x \in \text{dom } A_1$. Then the function $x_2(t) = TS_1(t)x$ is continuously differentiable for $t \geqslant 0$ and satisfies the equation

$$d/dt\, x_2(t) = TA_1S_1(t)x = A_2TS_1(t)x = A_2x_2(t).$$

Hence $x_2(t) = S_2(t)x_2(0) = S_2(t)Tx$. Now (i) follows from the fact that $\text{dom } A_1$ is dense in X_1. □

1.3.9 THEOREM. *Let* (H1), (H3) *be satisfied and let* $K \in L(Y,X)$. *Then the following statements hold.*

(i) *There exists a unique* C_o*-semigroup* $S_K(t) : X \to X$ *such that the equation*

$$S_K(t)\iota x = S(t)\iota x + \int_0^t S_K(t-s)KCS(s)x\,ds \tag{31}$$

holds for all $x \in X$ *and every* $t \geq 0$.

(ii) *Let* A_K *be the infinitesimal generator of* $S_K(t)$. *Then* $\operatorname{ran} \iota \subset \operatorname{dom} A_K$ *and*

$$A_K \iota x = A \iota x + KCx, \quad x \in X. \tag{32}$$

Moreover A_K *is the closure of its restriction to* $\operatorname{ran} \iota$.

(iii) *If* Y *is finite-dimensional, then* $\operatorname{dom} A_K = \operatorname{ran} \iota$.

(iv) *Let* $B \in L(U,X)$ *satisfy* (H2) *and let* $F \in L(X,U)$ *be given such that* $BF = KC \in L(X,X)$. *Moreover, let* $S_F(t) : X \to X$ *be the semigroup which was introduced in Theorem 1.3.7. Then*

$$S_K(t)\iota = \iota S_F(t). \tag{33}$$

Proof. (i) Recall that A^* satisfies (H1) (Remark 1.3.1 and Lemma 1.3.2) and that C^* satisfies (H2) (Lemma 1.3.5). Hence it follows from Theorem 1.3.7 that there exists a unique C_o-semigroup $S_K^*(t) : X^* \to X^*$ satisfying the equation

$$\iota^* S_K^*(t)x^* = \iota^* S^*(t)x^* + \int_0^t S^*(t-s)C^*K^*S_K^*(s)x^*ds$$

for every $x^* \in X^*$ and $t \geq 0$. It is easy to see that this equation is equivalent to (31).

(ii) By Theorem 1.3.7 (iii), the infinitesimal generator A_K^* of $S_K^*(t)$ is given by

$$\operatorname{dom} A_K^* = \{x^* \in X^* | A^*\iota^*x^* + C^*K^*x^* \in \operatorname{ran} \iota^*\},$$

$$\iota^* A_K^* x^* = A^*\iota^*x^* + C^*K^*x^*.$$

We show that A_K^* is the adjoint operator of $\tilde{A}_K : \operatorname{ran} \iota \to X$ which is defined by

$$\tilde{A}_K \iota x = A\iota x + KCx, \; x \in X.$$

For this purpose, let $x^*, w^* \in X^*$ be given. Then $x^* \in \operatorname{dom} \tilde{A}_K^*$ and $\tilde{A}_K^* x^* = w^*$ if and only if the following equation holds for every $x \in X$ (Lemma 1.3.2 (v))

$$\langle w^*, \iota x\rangle = \langle x^*, \tilde{A}_K \iota x\rangle = \langle A^*\iota^*x^* + C^*K^*x^*, x\rangle.$$

This is equivalent to $x^* \in \text{dom } A_K^*$ and $A_K^* x^* = w^*$. We conclude that the adjoint operator A_K of A_K^* is the closure of $\tilde{A}_K$.

(iii) Let Y be finite-dimensional. Then the operator $\tilde{A}_K$, defined in the proof of (ii), is closed. In order to prove this, let $x_n \in X$ and $x, w \in X$ be given such that

$$x = \lim_{n\to\infty} \iota x_n, \quad w = \lim_{n\to\infty} \tilde{A}_K \iota x_n.$$

Then we have to show that $x \in \text{ran } \iota$ (if this is shown, then it follows from (ii) that $\tilde{A}_K x = w$). First note that

$$M = \{x^* \in X^* \mid \langle x^*, KCx_n\rangle \text{ is a bounded sequence}\}$$

$$= K^{*-1} \{y^* \in Y^* \mid \langle y^*, Cx_n\rangle \text{ is a bounded sequence}\}$$

is a closed subspace of X^*, since $\dim Y^* < \infty$. Moreover, for every $x^* \in \text{dom } A^*$, the sequence

$$\langle x^*, KCx_n\rangle = \langle x^*, \tilde{A}_K \iota x_n\rangle - \langle A^* x^*, \iota x_n\rangle$$

is bounded, and dom A^* is dense in X^* since X is reflexive. This implies that $M = X^*$. By the uniform boundedness theorem, we obtain that KCx_n is a bounded sequence in X. Since ran K is finite dimensional, this sequence has a convergent subsequence KCx_{n_k}, $k \in \mathbb{N}$. Hence the sequence $A\iota x_{n_k} = \tilde{A}_K \iota x_{n_k} - KCx_{n_k}$ is also convergent. Since A is closed, this implies that $x \in \text{dom } A = \text{ran } \iota$.

(iv) Let $x \in \text{dom } A_F$. Then it follows from (ii) and Theorem 1.3.7 (iii) that $\iota x \in \text{dom } A_K$ and

$$A_K \iota x = A\iota x + KCx = A\iota x + BFx = \iota A_F x.$$

Hence (iv) follows from Lemma 1.3.8. □

DYNAMIC OBSERVATION

The main feature of the output injection semigroup $S_K(t)$ - defined in Theorem 1.3.9 - is that it gives rise to the design of a (full order) observer of system (21), given by

$$d/dt\, z(t) = A_K z(t) - K y(t), \; z(0) = z_o \in X. \tag{34}$$

This Cauchy problem in the state space X has to be understood in the sense of 'mild solutions' which means that a solution z(t) of (34) is given by

$$z(t) = S_K(t) z_o - \int_0^t S_K(t-s) K y(s) ds, \quad t \geq 0. \tag{35}$$

If the semigroup $S_K(t)$ is stable, then this equation is in fact an observer for system (21) in the state space X. In order to make this precise, we introduce the output operator of system (21), $C_T : X \to L^q([0,T];Y)$, by defining

$$[C_T \iota x](t) = CS(t)x, \; 0 \leqslant t \leqslant T, \quad x \in X. \tag{36}$$

It follows from hypothesis (H3) that this operator C_T is well defined on all of X and bounded on this domain.

Now suppose that the input $y(t)$ of the observer equation (34) is precisely the output of system (21) which means that $y(\cdot) = C_T x$ for some $x \in X$. Then it is easy to see that the 'error' $e(t) = z(t) - S(t)x$ of the observer (34) is in fact described by the semigroup $S_K(t)$ (in the case $x = \iota x$, $x \in X$, this follows from (35), (36) and (31), and in general from the continuous dependence of the solutions on the initial states).

The following compactness results will turn out to be useful for checking the stability of the perturbed semigroups.

1.3.10 <u>LEMMA</u>. *Let* (H1), (H2) *be satisfied, let* $F \in L(X,U)$ *be a compact operator, and let* $S_F(t)$ *be the semigroup which was introduced in Theorem 1.3.7. Then the operator* $S_F(t) - S(t) \in L(X)$ *is compact for every* $t \geq 0$.

<u>Proof</u>. It is easy to see that the function $t \to S_F^*(t)F^* \in L(U^*,X^*)$ is continuous with respect to the uniform operator topology. So is the function $t \to FS_F(t) \in L(X,U)$. Hence the operator which maps $x \in X$ into the function

$$FS_F(\cdot)x \in C([0,T];U) \subset L^p([0,T];U)$$

is compact (Arzela-Ascoli). Now the compactness of the operator $S_F(t) - S(t)$ follows from (H2) and equation (29). □

1.3.11 LEMMA. *Let* (H1), (H3) *be satisfied, let* $K \in L(Y,X)$ *be a compact operator, and let* $S_K(t)$ *be the semigroup which was introduced in Theorem 1.3.9. Then the operator* $S_K(t) - S(t) \in L(X)$ *is compact for every* $t \geq 0$.

REMARKS ON THE LITERATURE

Infinite-dimensional linear systems with unbounded input and output operators (in particular, partial differential equations with boundary control) have been studied in a similar framework, for instance, by Lions [88], Lions-Magenes [89], Curtain-Pritchard [25], Pollock-Pritchard [128], Ichikawa [57, 58]. Unbounded perturbation results can also be found in Dunford-Schwartz [37], Kato [68], Goldstein [39] and Pazy [127]. However, all these authors (except Ichikawa) assume either that the inequality in hypothesis (H3) is satisfied pointwise or that there are even stronger conditions on A and C (see, e.g., Goldstein [39], Pazy [127]). All these assumptions require a smoothing property for the semigroup S(t) which is not satisfied in the case of delay systems. Only Ichikawa [57, 58] makes assumptions analogous to (H2) and (H3), but the operators A and B (respectively A and C in his papers) are of a special form, and his results are not as detailed and precise as is needed for our purpose.

2 State space theory for neutral functional differential systems

In this chapter we develop a state space approach for linear neutral functional differential equations (NFDE) of the form

$$d/dt\left(x(t) - Mx_t\right) = Lx_t.$$

The main point of view in our theory of NFDEs is the consideration of two different state concepts which are actually dual to each other.

The 'classical' way of introducing the state of a functional differential equation (FDE) with finite delay is to specify an initial function of suitable length which describes the past history of the solution (compare Section 1.2). An alternative state concept can be obtained by regarding an additional forcing term as the initial state of the system. This idea is due to Miller [104]. It was first discovered by Burns and Herdman [17] that these two notions of the state of a delay equation are dual to each other. More precisely, the evolution of the second state concept (forcing terms) is described by the adjoint semigroup of the one which is associated with the transposed equation in terms of the original state concept (initial functions).

For retarded functional differential equations (RFDE) both state concepts can be treated in the produce space

$$M^p = \mathbb{R}^n \times L^p([-h,0];\mathbb{R}^n)$$

(see, e.g., Bernier-Manitius [11], Delfour [28]). However, for NFDEs it is convenient to study the two state concepts in different state spaces. If the 'classical' state concept is treated in the product space M^p (Burns-Herdman-Stech [18, 19]), then the dual state concept will be taken in the dual space $W^{-1,p} = W^{1,q^*}$ of the Sobolev space

$$W^{1,q} = W^{1,q}([-h,0];\mathbb{R}^n)$$

$(1 < p < \infty,\ 1/p + 1/q = 1)$. If the original state concept is defined in the state space $W^{1,p}$ (Henry [48]), then the appropriate state space for the dual state concept will turn out to be the product space M^p. These duality relations shed a new light on the relationship between the semigroup $S(t) : M^p \to M^p$ of Burns, Herdman and Stech and the semigroup $\mathcal{S}(t) : W^{1,p} \to W^{1,p}$ of Henry, superseding the well-known definition of $\mathcal{S}(t)$ as the restriction of $S(t)$ to the domain of its generator.

The relation between the two state concepts will be described by so-called structural operators. These extend the concept of structural operators for RFDEs which has been developed in Bernier-Manitius [11], Manitius [93] and Delfour-Manitius [29].

2.1 THE SEMIGROUP APPROACH

Consider the linear NFDE

$$d/dt \left(x(t) - Mx_t \right) = Lx_t \tag{1}$$

where $x(t) \in \mathbb{R}^n$ for $t \geq -h$ and $x_t : [-h,0] \to \mathbb{R}^n$ is defined by $x_t(\tau) = x(t+\tau)$ for $-h \leq \tau \leq 0$ $(0 < h < \infty)$. We assume that L and M are bounded linear functionals on $C = C([-h,0];\mathbb{R}^n)$ with values in $\mathbb{R}^n$. These can be represented by normalized functions $\eta, \mu : [-h,0] \to \mathbb{R}^{n \times n}$ of bounded variation in the following way

$$L\phi = \int_{-h}^{0} d\eta(\tau)\phi(\tau), \quad M\phi = \int_{-h}^{0} d\mu(\tau)\phi(\tau), \quad \phi \in C,$$

(compare Section 1.2).

In order to obtain existence and uniqueness for the solutions of (1), we will always assume that

$$-1 \notin \sigma(\lim_{\tau \uparrow 0} \mu(\tau)) \tag{2}$$

(compare Condition (1.14) in Section 1.2). A solution of (1) is a function $x \in L^p_{loc}([-h,\infty);\mathbb{R}^n)$ with the property that the expression

$$w(t) = x(t) - Mx_t, \quad t \geq 0,$$

is absolutely continuous and satisfies $\dot{w}(t) = Lx_t$ for almost every $t \geq 0$. This means precisely that the pair $w(t)$, $t \geq 0$, and $x(t)$, $t \geq -h$, is a solution (in the sense of Definition 1.2.2) of the following system of the form (1.13)

$$\Sigma \quad \boxed{\begin{aligned} \dot{w}(t) &= Lx_t \\ x(t) &= w(t) + Mx_t \end{aligned}}$$

Note that such a solution $x(t)$ of (1) may not become absolutely continuous with time - in contrast to the retarded case (the absolutely continuous component $w(t) = x(t) - Mx_t$ of system Σ should be interpreted only as an auxiliary variable).

THE SEMIGROUP IN THE STATE SPACE M^p

It follows from (2) that Σ satisfies the assumptions of Theorem 1.2.3. Hence Σ admits a unique solution for every initial condition:

$$w(0) = \phi^0, \ x(\tau) = \phi^1(\tau), \ -h \leqslant \tau < 0, \tag{3}$$

where $\phi = (\phi^0,\phi^1) \in M^p$. The corresponding semigroup $S(t) : M^p \to M^p$ associates with every initial state $\phi \in M^p$ the state

$$S(t)\phi = (w(t),x_t) \in M^p$$

of Σ at time $t \geqslant 0$. The infinitesimal generator of $S(t)$ is given by

$$\text{dom } A = \{\phi \in M^p \mid \phi^1 \in W^{1,p}, \ \phi^0 = \phi^1(0) - M\phi^1\}$$

$$A\phi = (L\phi^1, \dot{\phi}^1)$$

(Theorem 1.2.6). This semigroup has been introduced recently by Burns, Herdman and Stech [18, 19].

2.1.1 REMARKS

(i) The above concept of a solution to the NFDE (1) goes back to Hale and Meyer [43]. They have defined a continuous function $x(t)$, $t \geqslant -h$, to be a solution of (1) if $x(t) - Mx_t$ is continuously differentiable for $t \geqslant 0$ and satisfies (1). Moreover, they have shown that (1) admits a unique solution for every initial condition $x(\tau) = \phi(\tau)$, $-h \leqslant \tau \leqslant 0$, $\phi \in C$ (see also Theorem 1.2.3 (iv)).

(ii) The semigroup $S_C(t) : C \to C$ of Hale and Meyer maps every initial state $\phi \in C$ into the corresponding solution segment $S_C(t)\phi = x_t$ of (1) at time $t \geqslant 0$. Its infinitesimal generator is given by

$$\text{dom } A_C = \{\phi \in C | \dot{\phi} \in C, \quad \dot{\phi}(0) = L\phi + M\dot{\phi}\}$$

$$A_C\phi \quad = \dot{\phi}$$

(Hale-Meyer [43, Lemma 2]).

(iii) The semigroup $S_C(t) : C \to C$ can be regarded as a restriction of $S(t) : M^p \to M^p$. In this case we have to identify every $\phi \in C$ with the pair $(\phi(0) - M\phi, \phi) \in M^p$. Then C becomes a dense subspace of M^p which is invariant under $S(t)$.

THE SEMIGROUP IN THE STATE SPACE $W^{1,p}$

Sometimes it is not useful to allow solutions of (1) which are not absolutely continuous - in particular, if the output depends on the derivative of the solution. In this case we rewrite equation (1) in the following way

$$\Omega \qquad \boxed{\dot{x}(t) = Lx_t + M\dot{x}_t} \; .$$

It has been proved by Henry [48] that this equation admits a unique solution $x \in W^{1,p}_{loc}([-h,\infty);\mathbb{R}^n)$ for every initial condition

$$x(\tau) = \phi(\tau), \; -h \leqslant \tau \leqslant 0, \tag{4}$$

where $\phi \in W^{1,p}$ (compare Theorem 1.2.3 (v)). Moreover Henry [48] has introduced the C_o-semigroup $S(t) : W^{1,p} \to W^{1,p}$, which associates with every initial state $\phi \in W^{1,p}$ the corresponding solution segment $S(t)\phi = x_t \in W^{1,p}$ of Ω at time $t \geqslant 0$. We will see that this semigroup is nothing else than the restriction of S(t) to the domain of its generator. For this sake let us define the embedding $\iota : W^{1,p} \to M^p$ by

$$\iota\phi = (\phi(0) - M\phi, \phi) \in M^p, \ \phi \in W^{1,p}. \tag{5}$$

Then the range of ι is precisely the domain of A. Hence the operator A satisfies the hypothesis (H1) of Section 1.3 where $X = M^p$ and $X = W^{1,p}$.

Now let $\phi \in W^{1,p}$ be given and let w(t), x(t) be the unique solution of Σ corresponding to the initial state $\iota\phi \in M^p$. Then it follows from Theorem 1.2.3 (v) that $x \in W^{1,p}_{loc}([-h,\infty);\mathbb{R}^n)$. Hence $x(t) = x(t)$ satisfies Ω and (4). This can be written in the form $\mathcal{S}(t)\phi = x_t = [S(t)\iota\phi]^1$. Thus we have proved that

$$\iota\mathcal{S}(t) = S(t)\iota, \ t \geqslant 0, \tag{6}$$

(compare equation (1.22)). We conclude that the correspondence between the semigroups S(t) and $\mathcal{S}(t)$ is precisely the same as described in Section 1.3. In particular, it follows from Lemma 1.3.2 (i) that the generator of $\mathcal{S}(t)$ is given by

$$\text{dom } A = \{\phi \in W^{1,p} | \dot{\phi} \in W^{1,p}, \ \dot{\phi}(0) = L\phi + M\dot{\phi}\}$$

$$A\phi = \dot{\phi}$$

(see also Henry [48]).

We will see that the analog of the semigroup $\mathcal{S}(t)$ for the transposed system plays an important role in the description of the dual state concept for system Σ.

THE TRANSPOSED EQUATION

Transposition of matrices leads to the NFDE

$$d/dt\left(x(t) - M^T x_t\right) = L^T x_t \tag{7}$$

where the bounded, linear functionals L^T and M^T from C into $\mathbb{R}^n$ are given by

$$L^T = \int_{-h}^{0} d\eta^T(\tau)\psi(\tau), \ M^T\psi = \int_{-h}^{0} d\mu^T(\tau)\psi(\tau), \ \psi \in C.$$

We consider the 'classical' state concept of the NFDE (7) in the state spaces

M^q and $W^{1,q}$ $(1/p + 1/q = 1)$ in a way analogous to that used for the NFDE (1). The semigroup $S^T(t)$; $M^q \to M^q$ associates with every $\psi \in M^q$ the state $S^T(t)\psi = (z(t),x_t) \in M^q$ of system

$$\Sigma^T \qquad \boxed{\begin{aligned} \dot{z}(t) &= L^T x_t \\ x(t) &= z(t) + M^T x_t \end{aligned}}$$

at time $t \geqslant 0$, corresponding to the initial condition

$$z(0) = \psi^0, \ x(\tau) = \psi^1(\tau), \ -h \leqslant \tau < 0. \tag{8}$$

The infinitesimal generator of $S^T(t)$ is given by

$$\text{dom } A^T = \{\psi \in M^q | \psi^1 \in W^{1,q}, \ \psi^0 = \psi^1(0) - M^T\psi^1\}$$

$$A^T\psi \ = (L^T\psi^1, \ \dot{\psi}^1)$$

(Theorem 1.2.6). Correspondingly, we introduce the embedding $\iota^T : W^{1,q} \to M^q$ by defining

$$\iota^T\psi = (\psi(0) - M^T\psi, \psi) \in M^q, \ \psi \in W^{1,q}. \tag{9}$$

Again $\mathcal{S}^T(t) : W^{1,q} \to W^{1,q}$ is the unique semigroup which satisfies the equation

$$\iota^T\mathcal{S}^T(t) = S^T(t)\iota^T, \ t \geqslant 0. \tag{10}$$

Moreover, $\mathcal{S}^T(t)$ associates with every $\psi \in W^{1,q}$ the solution segment $\mathcal{S}^T(t)\psi = x_t \in W^{1,q}$ of system

$$\Omega^T \qquad \boxed{\dot{x}(t) = L^T x_t + M^T \dot{x}_t}$$

at time $t \geqslant 0$, corresponding to the initial condition

$$x(\tau) = \psi(\tau), \ -h \leqslant \tau \leqslant 0. \tag{11}$$

The infinitesimal generator of $\mathcal{S}^T(t)$ is given by

$$\text{dom } A^T = \{\psi \in W^{1,q} | \dot{\psi} \in W^{1,q},\ \dot{\psi}(0) = L^T\psi + M^T\dot{\psi}\}$$
$$A^T\psi = \dot{\psi}.$$

Summarizing this situation, we deal with the following four semigroups.

$$S(t) : M^p \to M^p \qquad\qquad S^T(t) : M^q \to M^q$$

$$\mathcal{S}(t) : W^{1,p} \to W^{1,p} \qquad\qquad \mathcal{S}^T(t) : W^{1,q} \to W^{1,q}$$

The semigroups on the left-hand side correspond to the NFDE (1) and the semigroups on the right-hand side to the transposed NFDE (7). On each side the semigroup below is the restriction of the upper semigroup to the domain of its generator. A diagonal relation will be introduced by the dual state concept.

THE DUAL STATE CONCEPT

For any type of delay system (FDEs, Volterra integral equations, integro-differential equations with infinite delays, difference equations) a dual state concept may be derived in the following way. The solution of the respective equation ($t > 0$) can be derived from the initial function ($t \leqslant 0$) in two steps. First, replace the initial function by an additional forcing term in the equation and, secondly, determine the solution of the equation which corresponds to this inhomogeneous term. The dual state concept is obtained by regarding the forcing term as the state of the system, rather than the solution segment (Miller).

An analogous procedure can be applied to the system Σ. For this purpose let us divide the right-hand side of each equation in Σ into two terms such that one of these depends only on the solution of Σ ($t > 0$) and the other only on the initial state $\phi \in M^p$ ($t \leqslant 0$).

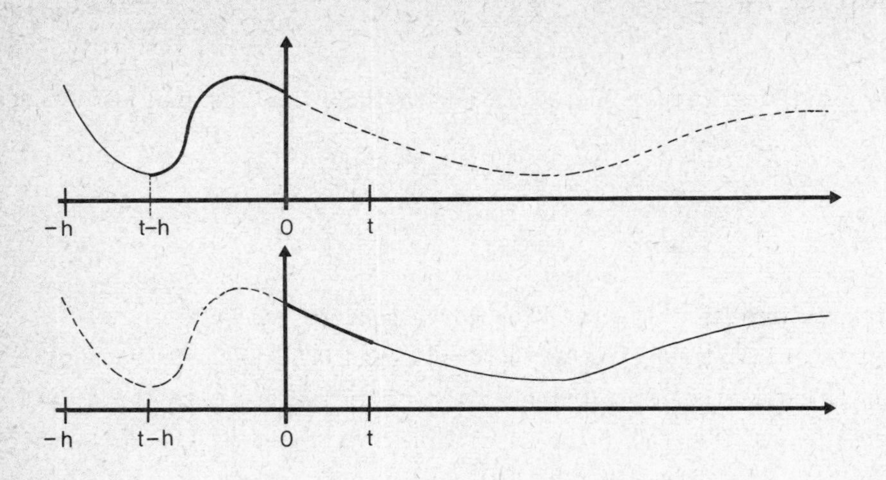

Figure 2

Then we obtain the following equations

$$\tilde{\Sigma} \quad \boxed{\begin{aligned} \dot{w}(t) &= \int_{-t}^{0} d\eta(\tau)x(t+\tau) + f^1(-t),\ w(0) = f^0, \\ x(t) &= w(t) + \int_{-t}^{0} d\mu(\tau)x(t+\tau) + f^2(-t),\ t \geqslant 0, \end{aligned}}$$

where the triple

$$f = (f^0, f^1, f^2) \in M^p = \mathbb{R}^n \times L^p([-h,0];\mathbb{R}^n) \times L^p([-h,0];\mathbb{R}^n)$$

is given by

$$f^0 = \phi^0 \tag{12.1}$$

$$f^1(-t) = \int_{-h}^{-t} d\eta(\tau)\phi^1(t+\tau),\ 0 \leqslant t \leqslant h, \tag{12.2}$$

$$f^2(-t) = \int_{-h}^{-t} d\mu(\tau)\phi^1(t+\tau),\ 0 \leqslant t \leqslant h, \tag{12.3}$$

$(f^1(-t) = f^2(-t) = 0$ for $t > h)$. By Remark 1.2.1 (ii), f^1 and f^2 are well-defined elements of $L^p([-h,0];\mathbb{R}^n)$, depending continuously on $\phi \in M^p$.

At first glance it seems natural to define the forcing term $f \in \mathcal{M}^p$ as the initial state of $\tilde{\Sigma}$, since the solution of $\tilde{\Sigma}$ depends only on $f \in \mathcal{M}^p$ rather than $\phi \in M^p$. However, it turns out that $\mathcal{M}^p$ is too large as a state space for $\tilde{\Sigma}$. In fact, different forcing terms might lead to the same solution $x(t)$ for $t \geqslant 0$ (recall that $w(t)$ is an auxiliary variable). More precisely, we will see that $x(t) = 0$ for $t \geqslant 0$ if and only if

$$\psi^T(0)f^0 + \int_{-h}^{0}\left(\psi^T(\tau)f^1(\tau) + \dot{\psi}^T(\tau)f^2(\tau)\right)d\tau = 0$$

for every $\psi \in W^{1,q}$ (Lemma 2.1.5). This suggests the introduction of the (bounded, linear) map $\pi : \mathcal{M}^p \to W^{-1,p}$, which associates with every $f \in \mathcal{M}^p$ the bounded, linear functional $\pi f \in W^{-1,p}$ on $W^{1,q}$, given by

$$\begin{aligned} &\langle\psi,\pi f\rangle_{W^{1,q},W^{-1,p}} \\ &= \psi^T(0)f^0 + \int_{-h}^{0} \psi^T(\tau)f^1(\tau)d\tau + \int_{-h}^{0} \dot{\psi}^T(\tau)f^2(\tau)d\tau \end{aligned} \tag{13}$$

for every $\psi \in W^{1,q}$. Then the forcing term $f \in \mathcal{M}^p$ is in the kernel of π if and only if the corresponding solution $x(t)$ of $\tilde{\Sigma}$ vanishes for $t \geqslant 0$. Motivated by this fact, we define the initial state of $\tilde{\Sigma}$ to be the bounded linear functional $\pi f \in W^{-1,p}$ which is represented by $f \in \mathcal{M}^p$ - rather than the forcing term f itself. This choice seems to represent the happy mean of a state space which carries no unnecessary burden but still contains all the information which is necessary in order to determine the solution of the system.

The corresponding state at time $t \geqslant 0$ can be obtained by applying a time shift to system $\tilde{\Sigma}$. The solution pair $w(t+s)$, $x(t+s)$, $s \geqslant 0$, of $\tilde{\Sigma}$ is determined by the forcing terms w^t, $x^t \in L^p([-h,0];\mathbb{R}^n)$ of the shifted equation

$$\begin{aligned} \dot{w}(t+s) &= \int_{-s}^{0} d\eta(\tau)x(t+s+\tau) + w^t(-s) \\ x(t+s) &= w(t+s) + \int_{-s}^{0} d\mu(\tau)x(t+s+\tau) + x^t(-s), \ s \geqslant 0. \end{aligned} \tag{14}$$

These forcing terms are given by

$$w^t(-s) = \int_{-s-t}^{-s} d\eta(\tau)x(t+s+\tau) + f^1(-s-t),\ 0 \leqslant s \leqslant h,$$
$$x^t(-s) = \int_{-s-t}^{-s} d\mu(\tau)x(t+s+\tau) + f^2(-s-t),\ 0 \leqslant s \leqslant h. \tag{15}$$

Now the state of $\tilde{\Sigma}$ at time $t \geqslant 0$ is the bounded linear functional

$$\pi(w(t),w^t,x^t) \in W^{-1,p}$$

on $W^{1,q}$. The evolution of this state is actually described by the semigroup $S^{T*}(t) : W^{-1,p} \to W^{-1,p}$. This is a consequence of Theorem 2.3.6 below.

2.1.2 COROLLARY. *Let* $f \in M^p$ *be given and let* $w(t)$, $x(t)$, $t \geqslant 0$, *be the corresponding solution of* $\tilde{\Sigma}$. *Moreover let* w^t *and* x^t *be defined by* (15). *Then* $\pi(w(t),w^t,x^t) = S^{T*}(t)\pi f$.

Let us now briefly discuss the question of whether or not a further restriction of the state space $W^{-1,p}$ of $\tilde{\Sigma}$ (to some invariant subspace of the semigroup $S^{T*}(t)$) might be useful. The reason for stressing this point is the fact that, in the retarded case ($\mu(\tau) \equiv 0$), the third component f^2 of f may be omitted and hence the produce space M^p is an appropriate state space for $\tilde{\Sigma}$. Note that this product space M^p can be embedded into $W^{-1,p}$ as a dense subspace via the map $\iota^{T*} : M^p \to W^{-1,p}$ (Remark 1.3.1 (iv)). This embedding associates with every pair $f = (f^0,f^1) \in M^p$ the bounded linear functional

$$\psi \to \langle \iota^T\psi, f\rangle_{M^q,M^p},\ \psi \in W^{1,q}.$$

2.1.3 REMARK. It follows from equation (10) that

$$S^{T*}(t)\iota^{T*} = \iota^{T*}S^{T*}(t),\ t \geqslant 0. \tag{16}$$

Moreover, by Lemma 1.3.2 (iv), we have dom A^{T*} = ran ι^{T*}. This means that the semigroup $S^{T*}(t) : M^p \to M^p$ represents the restriction of $S^{T*}(t) : W^{-1,p} \to W^{-1,p}$ to the domain of its generator.

In view of these facts it might be desirable to reduce the state space $W^{-1,p}$ of $\tilde{\Sigma}$ to the range of ι^{T*} which would lead to the semigroup $S^{T*}(t)$ on the state space M^p. However, this cannot be done directly since the bounded, linear functional $\pi f \in W^{-1,p}$ - arising from a forcing term $f \in M^p$ of $\tilde{\Sigma}$ which is given by (12) - will in general not be in the range of ι^{T*}.

Another possibility of a state space reduction for the system $\tilde{\Sigma}$ may be given through the use of the isomorphism

$$(\lambda I - A^{T*})\iota^{T*} : M^p \to W^{-1,p}$$

for some $\lambda \notin \sigma(A^T)$ (Remark 1.3.1 (ii)). This would again lead to the semigroup $S^{T*}(t)$ and the state space M^p (Lemma 1.3.2 (iii)). However, the price for such a somewhat artificial construction would be a more complicated relation between the two state concepts. Moreover, the Banach space $W^{-1,p}$ would still be needed as an intermediate step. Last, and not least, we are equally interested in the meaning of the different state spaces M^p and $W^{1,p}$ for the properties of the NFDE (1). This will come out through the above choice of $W^{-1,p}$ as a state space for $\tilde{\Sigma}$.

The desired restriction of the state space $W^{-1,p}$ for the dual state concept of the NFDE (1) to M^p can in fact be obtained in a direct way if we also restrict the state space M^p for the original state concept of the NFDE (1) to $W^{1,p}$. This restriction is represented by system Ω.

THE DUAL STATE CONCEPT FOR SYSTEM Ω

As above, we divide the right-hand side of Ω into two terms such that one of these depends only on the solution $x(t)$, $t > 0$, and the other only on the initial function $\phi \in W^{1,p}$. This procedure leads to the equation

$$\tilde{\Omega} \quad \boxed{\begin{aligned} &\dot{x}(t) = \int_{-t}^{0} d\eta(\tau)x(t+\tau) + \int_{-t}^{0} d\mu(\tau)\dot{x}(t+\tau) + f^1(-t), \; t \geq 0, \\ &x(0) = f^0, \end{aligned}}$$

where the pair $f = (f^0, f^1) \in M^p$ is given by

$$f^0 = \phi(0), \tag{17.1}$$

$$f^1(-t) = \int_{-h}^{-t} d\eta(\tau)\phi(t+\tau) + \int_{-h}^{-t} d\mu(\tau)\dot{\phi}(t+\tau), \; 0 \leq t \leq h. \tag{17.2}$$

2.1.4 <u>REMARK</u>. System $\tilde{\Omega}$ admits a unique solution $x \in W^{1,p}([0,\infty);\mathbb{R}^n)$ depending continuously on the forcing term $f \in M^p$. This can be seen by introducing the new variable $z(t) = \dot{x}(t)$ for $t \geq 0$ and defining $z(t) = x(t) = 0$ for $t < 0$. Then $\tilde{\Omega}$ is equivalent to the following system of the type (1.13)

$$\dot{x}(t) = z(t), \quad x(0) = f^0,$$

$$z(t) = Lx_t + Mz_t + f^1(-t), \quad t \geq 0.$$

Hence the above claim follows from Theorem 1.2.3.

The next result shows that $\tilde{\Omega}$ can in fact be regarded as a restriction of $\tilde{\Sigma}$ to the state space M^p.

2.1.5 LEMMA. *Let* $f \in M^p$ *and* $f \in M^p$ *be given. Moreover let* $w(t)$, $x(t)$ *be the unique solution of* $\tilde{\Sigma}$ *and* $x(t)$ *the unique solution of* $\tilde{\Omega}$. *Then the following statements are equivalent.*

(i) $x(t) = x(t)\ \forall\ t \geq 0$.

(ii) $\pi f = \iota^{T*} f$.

(iii) $$f^0 + \int_{-h}^{0} f^1(\tau)d\tau = \Big[I + \mu(-h)\Big]f^0 + \int_{-h}^{0} f^1(\tau)d\tau,$$

$$f^2(\sigma) + f^0 + \int_{\sigma}^{0} f^1(\tau)d\tau = \Big[I + \mu(\sigma)\Big]f^0 + \int_{\sigma}^{0} f^1(\tau)d\tau, \quad -h \leq \sigma \leq 0.$$

(iv) $$f^0 + \int_{-h}^{0} \Big(e^{\lambda\tau}f^1(\tau) + \lambda e^{\lambda\tau}f^2(\tau)\Big)d\tau$$

$$= \Big[I - M(e^{\lambda\cdot})\Big]f^0 + \int_{-h}^{0} e^{\lambda\tau}f^1(\tau)d\tau, \quad \forall\ \lambda \in \mathbb{C}.$$

Proof. First note that the equations

$$\langle\psi,\pi f\rangle = \psi^T(0)f^0 + \int_{-h}^{0} \psi^T(\tau)f^1(\tau)d\tau + \int_{-h}^{0} \dot{\psi}^T(\tau)f^2(\tau)d\tau$$

$$= \psi^T(-h)\Big(f^0 + \int_{-h}^{0} f^1(\tau)d\tau\Big) + \int_{-h}^{0} \dot{\psi}^T(\sigma)\Big(f^2(\sigma) + f^0 + \int_{\sigma}^{0} f^1(\tau)d\tau\Big)d\sigma \tag{18.1}$$

and

$$
\begin{aligned}
\langle \iota^T\psi, f\rangle &= \left(\psi(0) - M^T\psi\right)^T f^0 + \int_{-h}^{0} \psi^T(\tau) f^1(\tau)d\tau \\
&= \psi^T(0)f^0 + \psi^T(-h)\mu(-h)f^0 + \int_{-h}^{0} \dot{\psi}^T(\sigma)\mu(\sigma)d\sigma f^0 + \int_{-h}^{0} \psi^T(\tau)f^1(\tau)d\tau \\
&= \psi^T(-h)\left(f^0 + \mu(-h)f^0 + \int_{-h}^{0} f^1(\tau)d\tau\right) \\
&\quad + \int_{-h}^{0} \dot{\psi}^T(\sigma)\left(f^0 + \mu(\sigma)f^0 + \int_{\sigma}^{0} f^1(\tau)d\tau\right)d\sigma
\end{aligned} \tag{18.2}
$$

hold for every $\psi \in W^{1,q}$. This proves the equivalence of (ii) and (iii). Obviously (ii) implies (iv). Conversely, let (iv) be satisfied and apply (18) to $\psi(\tau) = e^{\lambda\tau}$, $-h \leqslant \tau \leqslant 0$. Then (iii) follows from the uniqueness of the Laplace transform. Hence it remains to prove that (i) is equivalent to (iii).

For this purpose let us introduce the functions

$$\tilde{x}(t) = x(t) - f^0, \; \tilde{f}(t) = w(t) + f^2(-t) - f^0 - \mu(-t)f^0,$$

for $t \geqslant 0$. Then the following equation holds

$$
\begin{aligned}
\tilde{x}(t) = x(t) - f^0 &= w(t) + f^2(-t) + \int_{-t}^{0} d\mu(\tau)x(t+\tau) - f^0 \\
&= \tilde{f}(t) + d\tilde{\mu} * \tilde{x}(t), \; t \geqslant 0.
\end{aligned}
$$

Now (i) is satisfied if and only if $x(t)$ is absolutely continuous for $t \geqslant 0$, $\tilde{x}(0) = 0$, and

$$
\begin{aligned}
\dot{\tilde{x}}(t) = \dot{x}(t) &= \int_{-t}^{0} d\eta(\tau)x(t+\tau) + \int_{-t}^{0} d\mu(\tau)\dot{x}(t+\tau) + f^1(-t) \\
&= \dot{w}(t) - f^1(-t) + f^1(-t) + d\tilde{\mu} * \dot{\tilde{x}}(t), \; t \geqslant 0.
\end{aligned}
$$

Equivalently, $\tilde{f}(t)$ is absolutely continuous for $t \geqslant 0$, $\tilde{f}(0) = 0$, and

$$\dot{\tilde{f}}(t) = \dot{w}(t) - f^1(-t) + f^1(-t), \; t \geqslant 0,$$

(Corollary 1.1.5). Since $w(0) = f^0$, this means that

$$\tilde{f}(t) = w(t) - f^0 - \int_{-t}^{0} f^1(\tau)d\tau + \int_{-t}^{0} f^1(\tau)d\tau, \; t \geqslant 0.$$

Finally, it follows from the definition of $\tilde{f}(t)$ that the latter is equivalent to (iii). □

Obviously, the solution $x(t)$ of $\tilde{\Omega}$ vanishes for $t \geq 0$ if and only if $f = 0$. Hence the product space M^p seems to be an appropriate choice for a state space of $\tilde{\Omega}$. The forcing term $f \in M^p$ will be regarded as the initial state of $\tilde{\Omega}$. The corresponding state at time $t \geq 0$ can again be derived from a time shift. The forcing term $x^t \in L^p([-h,0];\mathbb{R}^n)$ of the shifted equation

$$\dot{x}(t+s) = \int_{-s}^{0} d\eta(\tau)x(t+s+\tau) + \int_{-s}^{0} d\mu(\tau)\dot{x}(t+s+\tau) + x^t(-s), \quad s \geq 0, \quad (19)$$

is given by

$$x^t(-s) = \int_{-t-s}^{-s} d\eta(\tau)x(t+s+\tau) + \int_{-t-s}^{-s} d\mu(\tau)\dot{x}(t+s+\tau) + f^1(-t-s) \quad (20)$$

$(0 \leq s \leq h)$. The state of $\tilde{\Omega}$ at time $t \geq 0$ is defined to be the pair $(x(t),x^t) \in M^p$. The evolution of this state is described by the semigroup $S^{T*}(t) : M^p \to M^p$.

2.1.6 <u>COROLLARY</u>. *Let* $f \in M^p$ *be given and let* $x(t)$ *be the corresponding solution of* $\tilde{\Omega}$. *Moreover let* $x^t \in L^p([-h,0];\mathbb{R}^n)$ *be defined by* (20). *Then* $(x(t),x^t) = S^{T*}(t)f$.

<u>Proof</u>. Let $f \in M^p$ satisfy $\pi f = \iota^{T*}f$ and let $\pi(w(t),w^t,x^t) \in M^p$ be the corresponding state of $\tilde{\Sigma}$ defined by (15). Then it follows from Lemma 2.1.5 that $x(t) = x(t)$ and thus $\pi(w(t),w^t,x^t) = \iota^{T*}(x(t),x^t)$ for every $t \geq 0$. Hence, by Corollary 2.1.2 and (16),

$$\iota^{T*}(x(t),x^t) = S^{T*}(t)\pi f = S^{T*}(t)\iota^{T*}f = \iota^{T*}S^{T*}(t)f. \quad □$$

An explicit characterization of the infinitesimal generator A^{T*} of $S^{T*}(t)$ is given in the following proposition of which the proof is straightforward. The precise verification is left to the reader since this result will not be used later on.

2.1.7 <u>PROPOSITION</u>. *Let* $f, g \in M^p$ *be given. Then* $g \in \text{dom } A^{T*}$ *and* $A^{T*}g = f$ *if and only if the following equations hold:*

$$-\eta(-h)g^0 = [I + \mu(-h)]f^0 + \int_{-h}^{0} f^1(\tau)d\tau,$$

$$g^1(\sigma) - \eta(\sigma)g^0 = [I + \mu(\sigma)]f^0 + \int_{\sigma}^{0} f^1(\tau)d\tau, \ -h \leqslant \sigma \leqslant 0.$$

THE DUAL STATE CONCEPT FOR THE TRANSPOSED EQUATION

Precisely the same ideas as above can be applied to the systems Σ^T and Ω^T.

For the system Σ^T the dual state concept will be treated in the state space $W^{-1,q}$. More precisely, we obtain the equation

$$\tilde{\Sigma}^T \quad \boxed{\begin{aligned} \dot{z}(t) &= \int_{-t}^{0} d\eta^T(\tau)x(t+\tau) + g^1(-t), \ z(0) = g^0, \\ x(t) &= z(t) + \int_{-t}^{0} d\mu^T(\tau)x(t+\tau) + g^2(-t), \ t \geqslant 0, \end{aligned}}$$

where the triple $g = (g^0,g^1,g^2) \in M^q$ is given by expressions analogous to (12). The initial state of $\tilde{\Sigma}^T$ is the bounded, linear functional $\pi^T g \in W^{-1,q}$ which is given by

$$\begin{aligned} &\langle \pi^T g,\phi \rangle_{W^{-1,q},W^{1,p}} \\ &= g^{0^T}\phi(0) + \int_{-h}^{0} g^{1^T}(\tau)\phi(\tau)d\tau + \int_{-h}^{0} g^{2^T}(\tau)\dot{\phi}(\tau)d\tau \end{aligned} \tag{21}$$

for every $\phi \in W^{1,p}$. The corresponding state at time $t \geqslant 0$ is given by $\pi^T(z(t),z^t,x^t) \in W^{-1,q}$ where z^t, $x^t \in L^q([-h,0];\mathbb{R}^n)$ are the forcing terms of the shifted equation, i.e.

$$z^t(\sigma) = \int_{\sigma-t}^{\sigma} d\eta^T(\tau)x(t+\tau-\sigma) + g^1(\sigma-t), \ -h \leqslant \sigma \leqslant 0, \tag{22.1}$$

$$x^t(\sigma) = \int_{\sigma-t}^{\sigma} d\mu^T(\tau)x(t+\tau-\sigma) + g^2(\sigma-t), \ -h \leqslant \sigma \leqslant 0. \tag{22.2}$$

The evolution of this state is described by the semigroup $S^*(t)$ on $W^{-1,q}$, i.e.

$$\pi^T(z(t),z^t,x^t) = S^*(t)\pi^T g, \ t \geqslant 0. \tag{23}$$

The dual state concept of system Ω^T can be treated in the (restricted) state space M^q. We have the following equation

$$\tilde{\Omega}^T \quad \begin{array}{|l|} \hline \dot{x}(t) = \int_{-t}^{0} d\eta^T(\tau)x(t+\tau) + \int_{-t}^{0} d\mu^T(\tau)\dot{x}(t+\tau) + g^1(-t), \ t \geq 0, \\ x(0) = g^0. \\ \hline \end{array}$$

The initial state of $\tilde{\Omega}^T$ is the pair $g = (g^0, g^1) \in M^q$. The corresponding state at time $t \geq 0$ is given by $(x(t), x^t) \in M^q$ where $x^t \in L^q([-h,0];\mathbb{R}^n)$ is of the form

$$x^t(\sigma) = \int_{\sigma-t}^{\sigma} d\eta^T(\tau)x(t+\tau-\sigma) + \int_{\sigma-t}^{\sigma} d\mu^T(\tau)\dot{x}(t+\tau-\sigma) + g^1(\sigma-t) \tag{24}$$

$(-h \leq \sigma \leq 0)$. This expression can again be obtained by applying a time shift to $\tilde{\Omega}^T$. The evolution of the pair $(x(t), x^t) \in M^q$ is described by the semigroup $S^*(t)$, i.e.

$$(x(t), x^t) = S^*(t)g, \ t \geq 0. \tag{25}$$

As before, the system $\tilde{\Omega}^T$ represents the restriction of $\tilde{\Sigma}^T$ to the product space M^q via the embedding $\iota^* : M^q \to W^{-1,q}$ which associates with every $g \in M^q$ the bounded linear functional

$$\phi \to \langle g, \iota\phi \rangle_{M^q, M^p}, \ \phi \in W^{1,p}.$$

More precisely, we have the following relation as a consequence of equation (6)

$$\mathcal{S}^*(t)\iota^* = \iota^* S^*(t), \ t \geq 0. \tag{26}$$

Also, the analogue of Lemma 2.1.5 holds for the systems $\tilde{\Sigma}^T$ and $\tilde{\Omega}^T$.

The duality relation between the systems Σ and Ω^T can now be described by means of the following four semigroups

$$\begin{array}{ll} S(t) & \mathcal{S}^T(t) \\ \mathcal{S}^{T*}(t) & S^*(t) \end{array}$$

The semigroups on the left-hand side correspond to Σ and the semigroups on the right-hand side to Ω^T. On each side the upper semigroup describes the respective equation within the original state concept (initial functions) and the semigroup below describes that within the dual state concept (forcing terms). The diagonal relation is actually given by functional analytic duality theory.

In an analogous way the semigroups

$$\begin{matrix} \mathcal{S}(t) & S^T(t) \\ S^{T*}(t) & \mathcal{S}^*(t) \end{matrix}$$

correspond to the systems Ω and Σ^T.

In the next section we will clarify the relation between the semigroups $S(t)$ and $\mathcal{S}^{T*}(t)$, respectively between $\mathcal{S}(t)$ and $S^{T*}(t)$. This can be done by introducing so-called structural operators (Manitius).

REMARKS ON THE LITERATURE

RFDEs have been studied extensively in the product space M^p since the 'classical' paper of Borisovic and Turbabin [16]. The basic theory in this framework has been developed, for instance, by Delfour-Mitter [30], Banks-Burns [1, 2], Vinter [145], Bernier-Manitius [11], Delfour [26, 28] and Manitius [93]. In particular, the recent papers of Vinter [146], Delfour [27] and Delfour-Manitius [29] have shown that RFDEs can be treated in the product space in full generality. For the study of RFDEs in the state space C of continuous functions we refer to the book by Hale [42].

The work on NFDEs has been done mainly in the state spaces C (Hale-Meyer [43], Hale [42], Henry [49, 50]) and $W^{1,p}$ (Henry [48], Banks-Jacobs-Langenhop [6], Jakubczyk [62], Bartosiewicz [9, 10], O'Connor [109]). Recently Burns, Herdman and Stech [18, 19] have developed the basic ideas for the study of NFDEs in the product space M^p. Ito [59] has used these ideas for the study of the linear quadratic problem for NFDEs in the state space M^2.

The dual state concept was first introduced by Miller [104] for Volterra integro-differential equations with infinite delays. The corresponding duality result for the same class of systems has been shown by Burns and Herdman [17]. Diekmann [32] has applied these ideas to Volterra integral equations in the state space C. For RFDEs results in this direction can be found in

Manitius [93], Delfour-Manitius [29], Delfour [28] and Salamon [135] within the product space framework and - in a slightly different way - in Diekmann [33] in the state space C. In the earlier work on RFDEs (Hale [42], Henry [47]) and NFDEs (Hale-Meyer [43], Henry [48, 49], O'Connor [109]) these explicit duality results were masked by the concept of the hereditary product (see below).

2.2 THE STRUCTURAL OPERATORS

We have seen that the solution segment of a delay equation at the maximal delay time h can be derived from the initial function in two steps (see p. 41). These two operations can be expressed by the so-called 'structural operators' F and G. Roughly speaking, the operator F maps the initial function into the corresponding forcing term of the equation, and the operator G maps this forcing term into the corresponding solution segment at time h. Operators of this type provide a very useful tool for the state space description as well as the analysis of control and observation properties of delay systems.

The operator F was first introduced by Bernier and Manitius [11] for retarded systems in the product space M^p. Later, Manitius [93] introduced the operator G for the same class of systems in connection with the study of the completeness problem. Further results on the role of the structural operator F in the theory of RFDEs can be found in Delfour-Manitius [29]. Recently, Diekmann [33] has defined an analogue of the F-operator for retarded systems in the state space C.

For neutral systems an F-operator in the state space $W^{1,2}$ can be found in O'Connor [109]. However, that operator cannot be composed with an operator G in the sense indicated above. But the effectiveness of the structural operator approach lies in the relation of the two operators F and G. Therefore we will define the operator F in a different way.

For the description of system Σ we will introduce the structural operators $F : M^p \to W^{-1,p}$ and $G : W^{-1,p} \to M^p$ as follows. The operator F associates with every $\phi \in M^p$ the bounded, linear functional

$$F\phi = \pi f \in W^{-1,p} \ (f \in M^p \text{ given by (12)}) \tag{27}$$

which is represented by the corresponding forcing term $f \in M^p$ of system $\tilde{\Sigma}$. The operator $G : W^{-1,p} \to M^p$ is defined by the relation

$$G\pi f = (w(h), x_h) \in M^p, \; f \in \mathcal{M}^p, \tag{28}$$

where the pair $w(t)$, $x(t)$, $t \geq 0$, is the unique solution of $\tilde{\Sigma}$ corresponding to $f \in \mathcal{M}^p$.

2.2.1 LEMMA. *There is a unique bounded linear operator* $G : W^{-1,p} \to M^p$ *satisfying* (28) *for every* $f \in M^p$. *This operator is bijective.*

Proof. Let us introduce the operator $\tilde{G} : \mathcal{M}^p \to M^p$ which associates with every forcing term $f \in \mathcal{M}^p$ the corresponding solution segment $\tilde{G}f = (w(h), x_h) \in M^p$ of system $\tilde{\Sigma}$. This operator is obviously bounded and linear. Moreover it follows from Lemma 2.1.5 that $\ker \tilde{G} = \ker \pi$. Hence $\tilde{G}$ induces an injective operator $\tilde{\tilde{G}}$ from $\mathcal{M}^p/\ker \pi$ into M^p. Note that the map $[f] \to \pi f$ from $\mathcal{M}^p/\ker \pi$ onto $W^{-1,p}$ is an isomorphism. We conclude that there exists a unique bounded, linear, one-to-one map $G : W^{-1,p} \to M^p$ satisfying $G\pi f = \tilde{\tilde{G}}[f] = \tilde{G}f$ for every $f \in \mathcal{M}^p$.

It remains to prove that G is onto. For this sake let $\phi \in M^p$ be given and define

$$x(t) = \phi^1(t-h), \; w(t) = \phi^0 - \int_t^h [\int_{-s}^0 d\eta(\tau)\phi^1(s+\tau-h)] \, ds$$

for $0 \leq t \leq h$. Then $x(t)$ and $w(t)$ satisfy equation $\tilde{\Sigma}$ where $f^0 = w(0)$, $f^1 = 0$, and

$$f^2(-t) = x(t) - w(t) - \int_{-t}^0 d\mu(\tau)x(t+\tau), \; 0 \leq t \leq h. \quad \square$$

These two operators F and G have the following important properties (compare Bernier-Manitius [11], Manitius [93] for RFDEs).

2.2.2 THEOREM. *Let the operators* F *and* G *be defined as above. Then*

$$S(h) = GF \qquad S^T(h) = G^*F^*$$
$$S^{T*}(h) = FG \qquad S^*(h) = F^*G^* \tag{29}$$

and for every $t \geq 0$

$$FS(t) = \mathcal{S}^{T*}(t)F \quad F^*\mathcal{S}^T(t) = S^*(t)F^*$$
$$S(t)G = G\mathcal{S}^{T*}(t) \quad \mathcal{S}^T(t)G^* = G^*S^*(t) \tag{30}$$

Proof. The equations on the left-hand side of (29) follow directly from the definition of the operators F and G. The equations on the right-hand side can be obtained by taking the adjoint operators.

Now let $\phi \in M^p$ be given and let $w(t)$, $t \geq 0$, and $x(t)$, $t \geq -h$, be the corresponding solution pair of Σ, (3). Moreover let $f \in M^p$ be given by (12). Then $w(t)$ and $x(t)$ satisfy $\tilde{\Sigma}$ for $t \geq 0$. Hence $F(w(t),x_t) = \pi(w(t),w^t,x^t)$ for every $t \geq 0$ where w^t and x^t are defined by (15). By Corollary 2.1.2, this implies

$$FS(t)\phi = \pi(w(t),w^t,x^t) = \mathcal{S}^{T*}(t)\pi f = \mathcal{S}^{T*}(t)F\phi.$$

On the other hand, let $f \in M^p$ be given and let $w(t)$, $x(t)$, $t \geq 0$ be the corresponding solution pair of $\tilde{\Sigma}$. Moreover, let $(w(t),w^t,x^t) \in M^p$ be defined by (15). Then it follows from (14) and Corollary 2.1.2 that

$$G\mathcal{S}^{T*}(t)\pi f = G\pi(w(t),w^t,x^t) = (w(t+h),x_{t+h})$$
$$= S(t)(w(h),x_h) = S(t)G\pi f. \quad \square$$

The equations on the left-hand side of (29) and (30) may be illustrated by the commuting diagram below

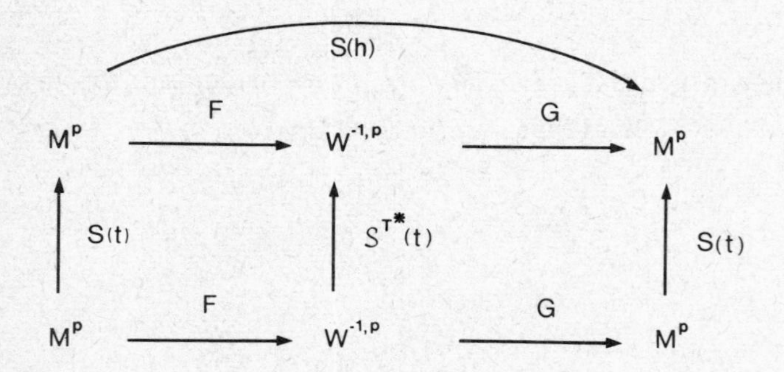

So far the equations on the right-hand side of (29) and (30) are obtained simply by dualizing the equations on the left-hand side. However, it is important not only to make use of these equations in a purely formal way but to understand their meaning. More precisely, we will see that these equations can be interpreted via the two state concepts of system Ω^T. For this purpose we have to show that $F^* : W^{1,q} \to M^q$ and $G^* : M^q \to W^{1,q}$ are the structural operators of system Ω^T. This means that they are associated with Ω^T in the same manner as the operators $F : W^{1,p} \to M^p$ and $G : M^p \to W^{1,p}$ (defined below) are associated with system Ω.

The operator $F : W^{1,p} \to M^p$ associates with every $\phi \in W^{1,p}$ the corresponding forcing term

$$F\phi = f \in M^p \text{ (f given by (17))} \tag{31}$$

of system $\tilde{\Omega}$. The operator $G : M^p \to W^{1,p}$ maps every $f \in M^p$ into the corresponding solution segment of $\tilde{\Omega}$ at time h, i.e.

$$Gf = x_h \in W^{1,p} \tag{32}$$

where $x(t)$, $t \geqslant 0$, is the unique solution of $\tilde{\Omega}$.

2.2.3 <u>THEOREM</u>. *Let the operators* F *and* G *be defined as above. Then*

$$\begin{aligned} S(h) &= GF \qquad & S^T(h) &= G^*F^* \\ S^{T*}(h) &= FG \qquad & S^*(h) &= F^*G^* \end{aligned} \tag{33}$$

and for every $t \geqslant 0$

$$\begin{aligned} FS(t) &= S^{T*}(t)F \qquad & F^*S^T(t) &= S^*(t)F^* \\ S(t)G &= GS^{T*}(t) \qquad & S^T(t)G^* &= G^*S^*(t). \end{aligned} \tag{34}$$

This theorem may be understood in two different ways. On the one hand it is the result analogous to Theorem 2.2.2 with Σ replaced by the (restricted) system Ω. On the other hand, Theorem 2.2.3 can be obtained by applying Theorem 2.2.2 to system Σ^T instead of Σ. For the second interpretation of Theorem 2.2.3 as a corollary of Theorem 2.2.2, we have to show that $F^* : M^q \to W^{-1,q}$ and $G^* : W^{-1,q} \to M^q$ are the structural operators of system Σ^T. To do this we need an explicit representation of the operator G. This can be given via the fundamental solution of the NFDE (1).

THE FUNDAMENTAL SOLUTION

The fundamental solution of a NFDE has been introduced by Hale and Meyer [43]. Further results can be found in Henry [49], Banks-Kent [7] and Kappel [67]. Recently, Ito [59] has applied these ideas to NFDEs in the state space M^2.

2.2.4 DEFINITION. *Let* $X : [0,\infty) \to \mathbb{R}^{n \times n}$ *be the unique function which is in* $NBV([0,T);\mathbb{R}^{n \times n})$ *for every* $T > 0$ *and satisfies the equation*

$$X = \rho + \tilde{\eta} * X + d\tilde{\mu} * X \tag{35}$$

where $\rho : [0,\infty) \to \mathbb{R}^{n \times n}$ *is defined by* $\rho(0) = 0$ *and* $\rho(t) = I$ *for* $t > 0$. *Moreover let us define*

$$W(t) = I - \int_0^t \eta(s-t)X(s)ds, \quad Z(t) = I - \int_0^t X(s)\eta(s-t)ds, \tag{36}$$

for $t \geqslant 0$ *and* $X(t) = W(t) = Z(t) = 0$ *for* $t < 0$. *Then the triple* $X(t)$, $W(t)$, $Z(t)$ *is said to be the fundamental solution of the* NFDE (1).

2.2.5 REMARKS

(i) If the tripe $X(t)$, $W(t)$, $Z(t)$ is the fundamental solution of (1), then $X = \rho + X * \tilde{\eta} + X * d\tilde{\mu}$ (Section 1.1). Hence the triple $X^T(t)$, $Z^T(t)$, $W^T(t)$ is the fundamental solution of the transposed NFDE (7).

(ii) For every $t > 0$ we have

$$W(t) = \rho(t) + \tilde{\eta} * X(t), \quad Z(t) = \rho(t) + X * \tilde{\eta}(t).$$

By Remark 1.1.1 (x), this implies that $W(t)$ and $Z(t)$ are absolutely continuous and that the following equations hold for almost every $t > 0$

$$\dot{W}(t) = d\tilde{\eta} * X(t), \quad \dot{Z}(t) = dX * \tilde{\eta}(t).$$

(iii) By (ii) and (35) respectively (i), we have

$$W(t) = X(t) - d\tilde{\mu} * X(t), \quad Z(t) = X(t) - X * d\tilde{\mu}(t)$$

for every $t > 0$. This implies

$$W - W * d\tilde{\mu} = X - d\tilde{\mu} * X - X * d\tilde{\mu} + d\tilde{\mu} * X * d\tilde{\mu}$$

$$= Z - d\tilde{\mu} * Z.$$

(iv) In general $W(t) \not\equiv Z(t)$. However, in the retarded case ($\mu(\tau) \equiv 0$) we have $W(t) = Z(t) = X(t)$ for every $t > 0$.

(v) It is well known that the function $t \to \operatorname{VAR}_{[0,t)} X$ is exponentially bounded as t goes to infinity (see, e.g., Hale [42] or Kappel [67]).

The notion *fundamental solution* for the triple $X(t)$, $W(t)$, $Z(t)$, as defined above, is justified by the following result.

2.2.6 PROPOSITION. *Let* $X(t)$, $W(t)$, $Z(t)$ *be the fundamental solution of the* NFDE (1). *Then the following statements hold.*

(i) *The unique solution pair* $w(t)$, $x(t)$, $t \geq 0$, *of* $\tilde{\Sigma}$, *corresponding to* $f \in M^p$, *is given by*

$$w(t) = W(t)f^0 + \int_0^t W(t-s)f^1(-s)ds + \int_0^t \dot{W}(t-s)f^2(-s)ds, \tag{37.1}$$

$$x(t) = X(t)f^0 + \int_0^t X(t-s)f^1(-s)ds + \int_0^t dX(s)f^2(s-t). \tag{37.2}$$

(ii) *The unique solution* $x(t)$, $t \geq 0$, *of* $\tilde{\Omega}$, *corresponding to* $f \in M^p$, *is given by*

$$x(t) = Z(t)f^0 + \int_0^t X(t-s)f^1(-s)ds. \tag{38}$$

Proof. (i) Let us define $\tilde{f}^i(t) = f^i(-t)$ for $t \geq 0$ and $i = 1,2$. Then, integrating the first equation in $\tilde{\Sigma}$, we obtain the following equivalent system of Volterra-Stieltjes integral equations

$$w = \tilde{\eta} * x + f^0 + \rho * \tilde{f}^1, \quad x = w + d\tilde{\mu} * x + \tilde{f}^2.$$

Hence $x \in L^p_{loc}([0,\infty);R^n)$ is the unique solution of

$$x = \tilde{\eta} * x + d\tilde{\mu} * x + f^0 + \rho * \tilde{f}^1 + \tilde{f}^2.$$

Since $X(t)$ is the fundamental solution of this equation in the sense of Definition 1.1.3, it follows from Theorem 1.1.4 that

$$x(t) = dX * [f^0 + \rho * \tilde{f}^1 + \tilde{f}^2](t)$$

$$= X(t)f^0 + X * \tilde{f}^1(t) + dX * \tilde{f}^2(t),\ t \geqslant 0.$$

By Remark 2.2.5 (ii), this implies

$$w(t) = \tilde{\eta} * x(t) + f^0 + \rho * \tilde{f}^1(t)$$

$$= \tilde{\eta} * X(t)f^0 + \tilde{\eta} * X * \tilde{f}^1(t) + \tilde{\eta} * dX * \tilde{f}^2(t)$$

$$+ f^0 + \rho * \tilde{f}^1(t)$$

$$= W(t)f^0 + W * \tilde{f}^1(t) + \dot{W} * \tilde{f}^2(t).$$

(ii) Let us define $\tilde{f}^1(t) = f^1(-t)$ for $t \geqslant 0$. Then, integrating $\tilde{\Omega}$, we obtain the following equivalent Volterra-Stieltjes integral equation

$$x = f^0 + \rho * \dot{x}$$

$$= f^0 + \rho * d\tilde{\eta} * x + d\tilde{\mu} * \rho * \dot{x} + \rho * \tilde{f}^1$$

$$= \tilde{\eta} * x + d\tilde{\mu} * x + f^0 - \tilde{\mu}f^0 + \rho * \tilde{f}^1.$$

Again, it follows from Theorem 1.1.4 that

$$x(t) = dX * [f^0 - \tilde{\mu}f^0 + \rho * \tilde{f}^1](t)$$

$$= X(t)f^0 - dX * \tilde{\mu}(t)f^0 + X * \tilde{f}^1(t)$$

$$= Z(t)f^0 + X * \tilde{f}^1(t). \qquad \square$$

The explicit representation of the operators $G : W^{-1,p} \to M^p$ and $G : M^p \to W^{1,p}$, given below, is a direct consequence of Proposition 2.2.6.

2.2.7 COROLLARY

(i) *If* $f \in M^p$, *then* $G\pi f \in M^p$ *is given by*

$$[G\pi f]^0 = W(h)f^0 + \int_{-h}^{0} W(h+\sigma)f^1(\sigma)d\sigma + \int_{-h}^{0} \dot{W}(h+\sigma)f^2(\sigma)d\sigma,$$

$$[G\pi f]^1(\tau) = X(h+\tau)f^0 + \int_{-h}^{0} X(h+\tau+\sigma)f^1(\sigma)d\sigma + \int_{0}^{h+\tau} dX(s)f^2(s-\tau-h).$$

(ii) *If* $f \in M^p$, *then* $Gf \in W^{1,p}$ *is given by*

$$[Gf](\tau) = Z(h+\tau)f^0 + \int_{-h}^{0} X(h+\tau+\sigma)f^1(\sigma)d\sigma.$$

Dualizing this result as well as the expressions (27) and (31), we obtain that the adjoint operators F*, G*, $\mathcal{F}^*$, and $\mathcal{G}^*$ are of the following form.

2.2.8 LEMMA.

(i) *If* $\psi \in W^{1,q}$, *then* $F^*\psi \in M^q$ *is given by*

$$[F^*\psi]^0 = \psi(0),$$

$$[F^*\psi]^1(\sigma) = \int_{-h}^{\sigma} d\eta^T(\tau)\psi(\tau-\sigma) + \int_{-h}^{\sigma} d\mu^T(\tau)\dot{\psi}(\tau-\sigma).$$

(ii) *If* $g \in M^q$, *then* $G^*g \in W^{1,q}$ *is given by*

$$[G^*g](\tau) = W^T(h+\tau)g^0 + \int_{-h}^{0} X^T(h+\tau+\sigma)g^1(\sigma)d\sigma.$$

(iii) *If* $\psi \in M^q$, *then* $\mathcal{F}^*\psi = \pi^T g \in W^{-1,q}$ *where* $g \in M^q$ *is given by*

$$g^0 = \psi^0,\ g^1(\sigma) = \int_{-h}^{\sigma} d\eta^T(\tau)\psi^1(\tau-\sigma),$$

$$g^2(\sigma) = \int_{-h}^{\sigma} d\mu^T(\tau)\psi^1(\tau-\sigma).$$

(iv) *If* $g \in M^q$, *then* $\mathcal{G}^*\pi^T g \in M^q$ *is given by*

$$[\mathcal{G}^*\pi^T g]^0 = Z^T(h)g^0 + \int_{-h}^{0} Z^T(h+\sigma)g^1(\sigma)d\sigma + \int_{-h}^{0} \dot{Z}^T(h+\sigma)g^2(\sigma)d\sigma,$$

$$[\mathcal{G}^*\pi^T g]^1(\tau) = X^T(h+\tau)g^0 + \int_{-h}^{0} X^T(h+\tau+\sigma)g^1(\sigma)d\sigma + \int_{0}^{h+\tau} dX^T(s)g^2(s-\tau-h).$$

Proof. It is enough to prove (i) and (ii).

(i) Let $\psi \in W^{1,q}$, $\phi \in M^p$ be given, and let $f \in M^p$ be defined by (12). Then $F\phi = \pi f \in W^{-1,p}$, and hence

$$\langle F^*\psi,\phi\rangle_{M^q,M^p} = \langle\psi,\pi f\rangle_{W^{1,q},W^{-1,p}}$$

$$= \psi^T(0)\phi^0 + \int_{-h}^{0}\int_{\tau}^{0} \psi^T(\tau-\sigma)d\eta(\tau)\phi^1(\sigma)d\sigma + \int_{-h}^{0}\int_{\tau}^{0} \dot\psi^T(\tau-\sigma)d\mu(\tau)\phi^1(\sigma)d\sigma$$

$$= \psi^T(0)\phi^0 + \int_{-h}^{0}\left(\int_{-h}^{\sigma} d\eta^T(\tau)\psi(\tau-\sigma) + \int_{-h}^{\sigma} d\mu^T(\tau)\dot\psi(\tau-\sigma)\right)^T \phi^1(\sigma)d\sigma.$$

(ii) Let $g \in M^q$ and $f \in M^p$ be given, and let $\psi \in W^{1,q}$ be defined by

$$\psi(\tau) = W^T(h+\tau)g^0 + \int_{-h}^{0} X^T(h+\tau+\sigma)g^1(\sigma)d\sigma, \ -h \leqslant \tau \leqslant 0.$$

Then it follows from Remark 1.1.1 (x) that

$$\dot\psi(\tau) = \dot W^T(h+\tau)g^0 + \int_{0}^{h+\tau} dX^T(s)g^1(s-\tau-h), \ -h \leqslant \tau \leqslant 0.$$

By Corollary 2.2.7, this implies

$$\langle G^*g,\pi f\rangle_{W^{1,q},W^{-1,p}} = \langle g,G\pi f\rangle_{M^q,M^p}$$

$$= g^{0^T} W(h)f^0 + \int_{-h}^{0} g^{0^T} W(h+\sigma)f^1(\sigma)d\sigma + \int_{-h}^{0} g^{0^T}\dot W(h+\sigma)f^2(\sigma)d\sigma$$

$$+ \int_{-h}^{0} g^{1^T}(\tau)X(h+\tau)d\tau f^0 + \int_{-h}^{0}\int_{-h}^{0} g^{1^T}(\tau)X(h+\tau+\sigma)f^1(\sigma)d\sigma d\tau$$

$$+ \int_{0}^{h}\int_{0}^{h-s} g^{1^T}(-t)dX(s)f^2(s+t)-h)dt$$

$$= \psi^T(0)f^0 + \int_{-h}^{0} \psi^T(\tau)f^1(\tau)d\tau + \int_{-h}^{0} \dot\psi^T(\tau)f^2(\tau)d\tau. \qquad \square$$

Now recall that the solutions of $\tilde\Sigma^T$ and $\tilde\Omega^T$ can be represented by means of the fundamental solution $X^T(t)$, $Z^T(t)$, $W^T(t)$ of the transposed NFDE (7) (Proposition 2.2.6). Hence the expressions in Lemma 2.2.8 (ii) and (iv) show that

$$G^*g = x_h \in W^{1,q}, \ g \in M^q,$$

where $x(t)$, $t \geq 0$, is the unique solution of $\tilde{\Omega}^T$, and that

$$G^* \pi^T g = (z(h), x_h) \in M^q, \quad g \in M^q,$$

where $z(t)$, $x(t)$, $t \geq 0$, is the unique solution pair of $\tilde{\Sigma}^T$. We conclude that the operators $F^* : W^{1,q} \to M^q$ and $G^* : M^q \to W^{1,q}$ (respectively $F^* : M^q \to W^{-1,q}$ and $G^* : W^{-1,q} \to M^q$) are in fact the structural operators associated with system Ω^T (respectively system Σ^T).

The following result shows that the operators F and F are more or less the same. More precisely, the operator $F : W^{1,p} \to M^p$ is the restriction of $F : M^p \to W^{-1,p}$ via the embeddings $\iota : W^{1,p} \to M^p$ and $\iota^{T*} : M^p \to W^{-1,p}$. By analogy, $G : M^p \to W^{1,p}$ represents the restriction of $G : W^{-1,p} \to M^p$ to ran ι^{T*}.

2.2.9 <u>LEMMA</u>

$$F\iota = \iota^{T*} F \qquad F^*\iota^T = \iota^* F^*$$

$$G\iota^{T*} = \iota G \qquad G^*\iota^* = \iota^T G^*$$

<u>Proof</u>. First let $\phi \in W^{1,p}$ and $\psi \in W^{1,q}$. Then

$$\langle \psi, F\iota\phi \rangle_{W^{1,q}, W^{-1,p}}$$

$$= \psi^T(0)\Big(\phi(0) - M\phi\Big) + \int_{-h}^{0} \int_{\tau}^{0} \psi^T(\sigma) d\eta(\tau) \phi(\tau-\sigma) d\sigma$$

$$+ \int_{-h}^{0} \int_{\tau}^{0} \dot{\psi}^T(\sigma) d\mu(\tau) \phi(\tau-\sigma) d\sigma$$

$$= \psi^T(0)\Big(\phi(0) - M\phi\Big) + \int_{-h}^{0} \int_{\tau}^{0} \psi^T(\sigma) d\eta(\tau) \phi(\tau-\sigma) d\sigma$$

$$+ \int_{-h}^{0} \left\{ \Big[\psi^T(\sigma) d\mu(\tau) \phi(\tau-\sigma) \Big]_{\sigma=\tau}^{\sigma=0} + \int_{\tau}^{0} \psi^T(\sigma) d\mu(\tau) \dot{\phi}(\tau-\sigma) d\sigma \right\}$$

$$= \Big(\psi(0) - M^T\psi\Big)^T \phi(0) + \int_{-h}^{0} \int_{\tau}^{0} \psi^T(\tau-\sigma) d\eta(\tau) \phi(\sigma) d\sigma$$

$$+ \int_{-h}^{0} \int_{\tau}^{0} \psi^T(\tau-\sigma)d\mu(\tau)\dot{\phi}(\sigma)d\sigma$$

$$= \langle \iota^T\psi, F\phi\rangle_{M^q,M^p}$$

$$= \langle \psi, \iota^{T*}F\phi\rangle_{W^{1,q},W^{-1,p}}.$$

Secondly, let $f \in M^p$, $g \in M^q$ be given, and define $\phi := Gf \in W^{1,p}$ and $\psi = G^*g \in W^{1,q}$. Then it follows from Corollary 2.2.7 and Lemma 2.2.8 that

$$\phi(\tau) = Z(h+\tau)f^0 + \int_{-h}^{0} X(h+\tau+\sigma)f^1(\sigma)d\sigma, \ -h \leqslant \tau \leqslant 0,$$

$$\psi(\sigma) = W^T(h+\sigma)g^0 + \int_{-h}^{0} X^T(h+\tau+\sigma)g^1(\tau)d\tau, \ -h \leqslant \sigma \leqslant 0.$$

By Remark 2.2.5 (ii), (iii), this implies

$$\langle g, \iota Gf\rangle_{M^q,M^p} = g^{0^T}\Big(\phi(0) - M\phi\Big) + \int_{-h}^{0} g^{1^T}(\tau)\phi(\tau)d\tau$$

$$= g^{0^T}\Big[Z(h) - \int_{-h}^{0} d\mu(\tau)Z(h+\tau)\Big]f^0$$

$$+ g^{0^T} \int_{-h}^{0}\Big[X(h+\sigma) - \int_{-h}^{0} d\mu(\tau)X(h+\tau+\sigma)\Big]f^1(\sigma)d\sigma$$

$$+ \int_{-h}^{0} g^{1^T}(\tau)Z(h+\tau)f^0 d\tau + \int_{-h}^{0}\int_{-h}^{0} g^{1^T}(\tau)X(h+\tau+\sigma)f^1(\sigma)d\sigma d\tau$$

$$= g^{0^T}\Big[W(h) - \int_{-h}^{0} W(h+\tau)d\mu(\tau)\Big]f^0 + \int_{-h}^{0} g^{0^T} W(h+\sigma)f^1(\sigma)d\sigma$$

$$+ \int_{-h}^{0} g^{1^T}(\tau)\Big[X(h+\tau) - \int_{-h}^{0} X(h+\tau+\sigma)d\mu(\sigma)\Big]d\tau f^0$$

$$+ \int_{-h}^{0}\int_{-h}^{0} g^{1^T}(\tau)X(h+\tau+\sigma)d\tau \ f^1(\sigma)d\sigma$$

$$= \left(\psi(0) - M^T\psi\right)^T f^0 + \int_{-h}^{0} \psi^T(\sigma) f^1(\sigma) d\sigma$$

$$= \langle \iota^T \psi, f \rangle_{M^q, M^p}$$

$$= \langle \iota^T G^* g, f \rangle_{M^q, M^p}$$

$$= \langle g, G \iota^{T*} f \rangle_{M^q, M^p}. \qquad \square$$

2.2.10 REMARKS

(i) For retarded systems ($\mu(\tau) \equiv 0$) the range of the operator $F : M^p \to W^{-1,p}$ is already contained in ran ι^{T*}. Correspondingly $F\phi$ depends only on $\iota\phi = (\phi(0),\phi) \in M^p$ for every $\phi \in W^{1,p}$. We conclude that there exists a unique operator $\bar{F} : M^p \to M^p$ such that

$$\bar{F}\iota = F, \quad \iota^{T*}\bar{F} = F. \tag{39}$$

The operator $\bar{F}$ maps every $\phi \in M^p$ into the pair $(f^0, f^1) \in M^p$ which is given by (12.1) and (12.2). This is precisely the F-operator which was introduced by Bernier and Manitius [11].

The G-operator of Manitius [93] is given by

$$\bar{G} = \iota G = G\iota^{T*} \in L(M^p). \tag{40}$$

These two operators make the following diagram commute

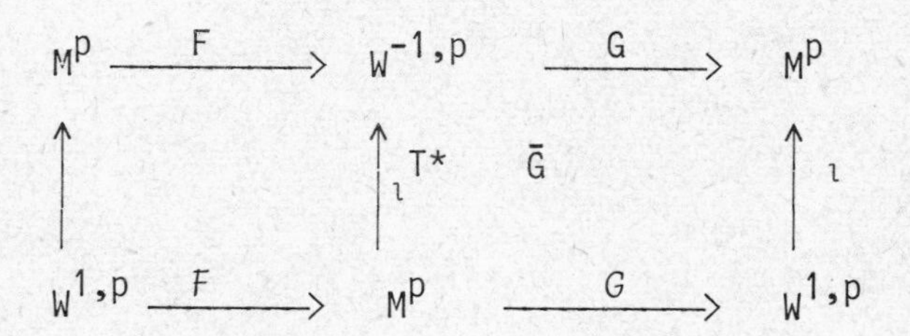

The existence of these operators is the reason why the dual state concept for RFDEs can also be treated in the product space M^p.

(ii) The operator $F = \iota^{T*}F : W^{1,p} \to W^{-1,p}$ is the structural operator which was introduced by O'Connor [109] for the case p = 2 (in [109] the dual space $W^{-1,2}$ is identified with $W^{1,2}$). This operator induces a bilinear form between $W^{1,q}$ and $W^{1,p}$ given by

$$<<\psi,\phi>> = <\psi,F\iota\phi>_{W^{1,q},W^{-1,p}} = <\iota^T\psi,F\phi>_{M^q,M^p}$$

$$= \psi^T(0)\Big(\phi(0) - M\phi\Big) + \int_{-h}^{0}\int_{\tau}^{0} \psi^T(\tau-\sigma)d\eta(\tau)\phi(\sigma)d\sigma$$

$$+ \int_{-h}^{0}\int_{\tau}^{0} \dot{\psi}^T(\tau-\sigma)d\mu(\tau)\phi(\sigma)d\sigma$$

$$= \Big(\psi(0) - M^T\psi\Big)^T\phi(0) + \int_{-h}^{0}\int_{\tau}^{0} \psi^T(\tau-\sigma)d\eta(\tau)\phi(\sigma)d\sigma$$

$$+ \int_{-h}^{0}\int_{\tau}^{0} {}^T(\tau-\sigma)d\mu(\tau)\dot{\phi}(\sigma)d\sigma$$

for $\phi \in W^{1,p}$ and $\psi \in W^{1,q}$. This so-called *hereditary product* has been introduced by Hale and Meyer [43] in the state space C. Further results in the state space $W^{1,2}$ can be found in Henry [48] and O'Connor [109].

(iii) We extend the above bilinear form to the case that either ϕ or ψ is in the product space. For $\psi \in M^q$ and $\phi \in W^{1,p}$ we define

$$<<\psi,\phi>> = <\psi,F\phi>_{M^q,M^p}$$

$$= \psi^{0^T}\phi(0) + \int_{-h}^{0}\int_{\tau}^{0} \psi^{1^T}(\tau-\sigma)d\eta(\tau)\phi(\sigma)d\sigma$$

$$+ \int_{-h}^{0}\int_{\tau}^{0} \psi^{1^T}(\tau-\sigma)d\mu(\tau)\dot{\phi}(\sigma)d\sigma$$

and for $\psi \in W^{1,q}$, $\phi \in M^p$

$$<<\psi,\phi>> = <\psi,F\phi>_{W^{1,q},W^{-1,p}}$$

$$= \psi^T(0)\phi^0 + \int_{-h}^{0}\int_{\tau}^{0} \psi^T(\tau-\sigma)d\eta(\tau)\phi^1(\sigma)d\sigma$$

$$+ \int_{-h}^{0}\int_{\tau}^{0} \dot{\psi}^T(\tau-\sigma)d\mu(\tau)\phi^1(\sigma)d\sigma.$$

With these definitions we have $\langle\langle\psi,\iota\phi\rangle\rangle = \langle\langle\iota^T\psi,\phi\rangle\rangle$ for every $\phi \in W^{1,p}$ and $\psi \in W^{1,q}$.

Finally, let us summarize the results of this section. This will be done by the two commuting diagrams below. The first diagram describes the relation between the various operators which are associated with the NFDE (1).

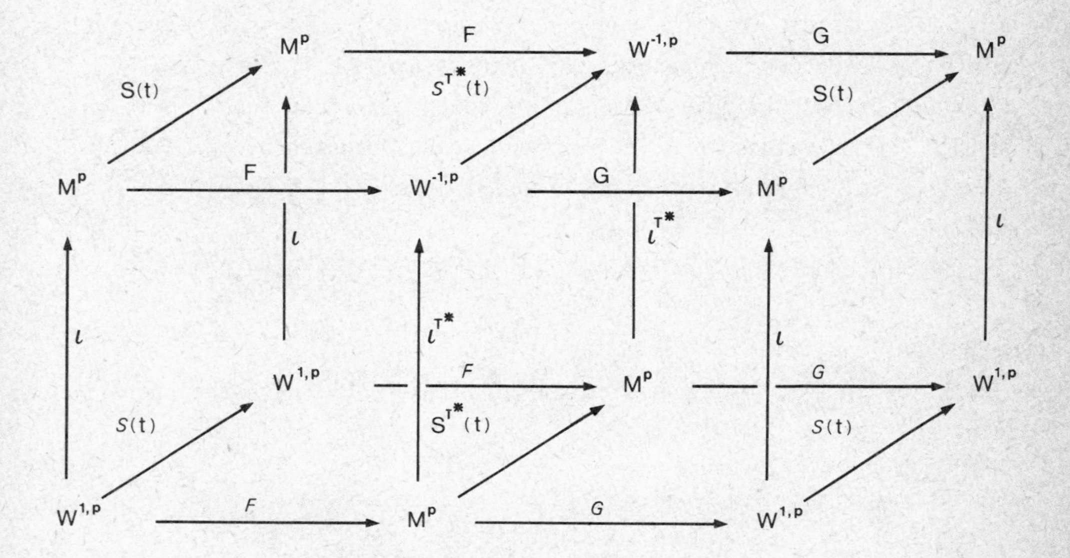

The adjoint diagram corresponds to the transposed NFDE (7).

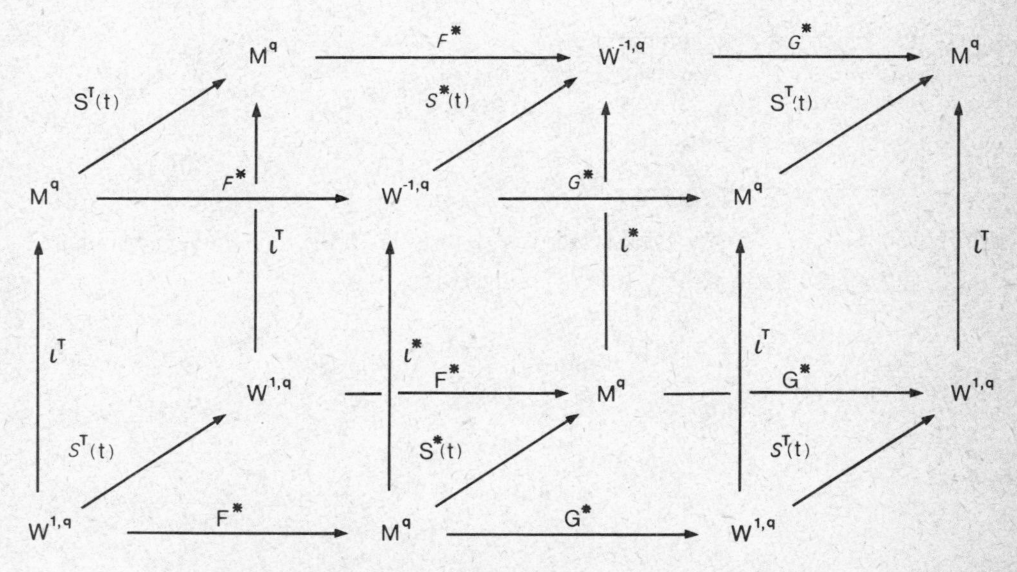

2.3 CONTROL AND OBSERVATION

In this section we develop a state space theory for the following control system which is governed by a linear NFDE having general delays in the input variables

$$d/dt\left(x(t) - Mx_t - \Gamma u_t\right) = Lx_t + Bu_t. \tag{41}$$

We will always assume that the control function $u(t) \in \mathbb{R}^m$ is locally p-times integrable. As before, u_t denotes the input segment $u_t(\tau) = u(t+\tau)$, $-h \leqslant \tau \leqslant 0$. Correspondingly, B and Γ are bounded, linear functionals on $C([-h,0];\mathbb{R}^m)$ with values in $\mathbb{R}^n$. These can be represented by normalized $n \times m$-matrix valued functions β and γ on the interval $[-h,0]$, i.e.

$$B\xi = \int_{-h}^{0} d\beta(\tau)\xi(\tau), \quad \Gamma\xi = \int_{-h}^{0} d\gamma(\tau)\xi(\tau), \quad \xi \in C([-h,0];\mathbb{R}^m),$$

(compare Section 1.2).

A solution of (41) is a function $x \in L^p_{loc}([-h,\infty);\mathbb{R}^n)$ with the property that the expression

$$w(t) = x(t) - Mx_t - \Gamma u_t, \quad t \geqslant 0,$$

is absolutely continuous and satisfies $\dot{w}(t) = Lx_t + Bu_t$ for almost every $t \geqslant 0$. This means that the pair $w(t)$, $x(t)$ is a solution of the following system of the form (1.13)

$$\Sigma \qquad \boxed{\begin{aligned} \dot{w}(t) &= Lx_t + Bu_t \\ x(t) &= w(t) + Mx_t + \quad u_t \end{aligned}}$$

(Definition 1.2.2). This system admits a unique solution for every input $u \in L^p_{loc}([0,\infty);\mathbb{R}^m)$ and every initial condition

$$w(0) = \phi^0, \quad x(\tau) = \phi^1(\tau), \tag{42.1}$$

$$u(\tau) = \xi(\tau), \quad -h \leqslant \tau \leqslant 0, \tag{42.2}$$

where $\phi \in M^p$ and $\xi \in L^p = L^p([-h,0];\mathbb{R}^m)$ (Theorem 1.2.3).

Let us first consider the simple case that there is no delay and no derivative in the input which means

$$B\xi = B_0\xi(0), \ \Gamma\xi = 0 \tag{43}$$

for $\xi \in C([-h,0];\mathbb{R}^m)$ $(B_0 \in \mathbb{R}^{n\times m})$. In this case it follows from Theorem 1.2.5 that the state $(w(t),x_t) \in M^p$ of Σ, (42) at time $t \geqslant 0$ is given by

$$(w(t),x_t) = S(t)\phi + \int_0^t S(t-s)(B_0u(s),0)ds \tag{44}$$

(see also Burns-Herdman-Stech [19, Theorem 3.1]). Clearly, the initial condition (42.2) on the input is, in this situation, not necessary in order to derive a solution of Σ, (43). This is still the case if $\Gamma\xi$ depends only on $\xi(0)$. However, then the evolution of $(w(t),x_t) \in M^p$ can no longer be described by an equation of type (44) through an input operator with values in M^p. In general, the initial condition (42.2) is really necessary in order to obtain a solution of Σ. There is a proper dependence of the solution on the past values of the input. This suggests the choice of the product space

$$M^p \times L^p$$

as a state space for Σ and the inclusion of the input segment u_t in the state of the system. Again, the evolution of the triple $(w(t),x_t,u_t) \in M^p \times L^p$ cannot be described by means of an input operator with values in $M^p \times L^p$. Such a description can be given, if we extend the state space. This extension is available through a restriction of the dual space (Remark 1.3.1 and Lemma 1.3.2). However, this procedure would involve computations with an explicit representation of the adjoint operator. In order to avoid these difficulties, we present an evolution equation approach for system Σ only within the dual state concept.

The dual state concept for system Σ can be introduced in precisely the same manner as in Section 2.1. Replacing the initial functions ϕ^1 and ξ by forcing terms, we obtain the equation

$$\tilde{\Sigma} \qquad \begin{array}{|l|} \hline \dot{w}(t) = \int_{-t}^{0} d\eta(\tau)x(t+\tau) + \int_{-t}^{0} d\beta(\tau)u(t+\tau) + f^1(-t) \\ \\ x(t) = w(t) + \int_{-t}^{0} d\mu(\tau)x(t+\tau) + \int_{-t}^{0} d\gamma(\tau)u(t+\tau) + f^2(-t) \\ \\ w(0) = f^0 \\ \hline \end{array}$$

where $f = (f^0, f^1, f^2) \in M^p$ is given by

$$f^0 = \phi^0, \tag{45.1}$$

$$f^1(\sigma) = \int_{-h}^{\sigma} d\eta(\tau)\phi^1(\tau-\sigma) + \int_{-h}^{\sigma} d\beta(\tau)\xi(\tau-\sigma), \ -h \leqslant \sigma \leqslant 0, \tag{45.2}$$

$$f^2(\sigma) = \int_{-h}^{\sigma} d\mu(\tau)\phi^1(\tau-\sigma) + \int_{-h}^{\sigma} d\gamma(\tau)\xi(\tau-\sigma), \ -h \leqslant \sigma \leqslant 0. \tag{45.3}$$

The initial state of $\tilde{\Sigma}$ is the bounded linear functional $\pi f \in W^{-1,p}$. The corresponding state at time $t \geqslant 0$ is given by $\pi(w(t), w^t, x^t) \in W^{-1,p}$ where w^t, $x^t \in L^p([-h,0];\mathbb{R}^n)$ are the forcing terms of the shifted equation $\tilde{\Sigma}$. These are of the following form

$$\begin{aligned} w^t(\sigma) &= \int_{\sigma-t}^{\sigma} d\eta(\tau)x(t+\tau-\sigma) + \int_{\sigma-t}^{\sigma} d\beta(\tau)u(t+\tau-\sigma) + f^1(\sigma-t), \\ x^t(\sigma) &= \int_{\sigma-t}^{\sigma} d\mu(\tau)x(t+\tau-\sigma) + \int_{\sigma-t}^{\sigma} d\gamma(\tau)u(t+\tau-\sigma) + f^2(\sigma-t) \end{aligned} \tag{46}$$

$(-h \leqslant \sigma \leqslant 0)$. We will see that this state of $\tilde{\Sigma}$ can be described by a variation-of-constants formula in the Banach space $W^{-1,p}$.

Now we want to study the NFDE (41) in the state space $W^{1,p} \times L^p$ within the original state concept (initial functions) and in the state space M^p within the dual state concept (forcing terms). This is only possible if, for every forced motion of system Σ with zero initial state, the pair $(w(t), x_t) \in M^p$ is in the range of ι. In order to ensure this we have to assume that $\Gamma = 0$, which means that there are no derivatives in the input.

THE RESTRICTED STATE SPACE ($\Gamma = 0$)

In the case $\Gamma = 0$ we may rewrite equation (41), respectively system Σ, in the following way

$$\Omega \qquad \boxed{\dot{x}(t) = Lx_t + M\dot{x}_t + Bu_t} \quad .$$

A solution $x(t)$, $t \geq -h$, of this system has to be absolutely continuous with L^p-derivative on every compact interval $[-h,T]$. Such a solution exists for every input $u \in L^p_{loc}([0,\infty);\mathbb{R}^m)$ and every initial condition

$$x(\tau) = \phi(\tau), \ -h \leq \tau \leq 0, \tag{47.1}$$

$$u(\tau) = \xi(\tau), \ -h \leq \tau < 0, \tag{47.2}$$

where $\phi \in W^{1,p}$ and $\xi \in L^p$ (compare Remark 2.1.4). This suggests the choice of the produce space

$$W^{1,p} \times L^p$$

as a state space for system Ω.

2.3.1 <u>REMARK</u>. In the case $\Gamma = 0$, system Ω represents the restriction of system Σ to the subspace ran $\iota \subset M^p$. In fact, let $w(t)$, $x(t)$ be the unique solution of Σ corresponding to the initial state $(\iota\phi,\xi)$, $\phi \in W^{1,p}$, $\xi \in L^p$, and to the input $u \in L^p_{loc}([0,\infty);\mathbb{R}^m)$. Then it follows from Theorem 1.2.3 (v) that $x(t)$ is absolutely continuous for $t \geq -h$ with locally p-times integrable derivative. Under this condition it is easy to check that $x(t)$ satisfies Ω.

Motivated by the same arguments as in the case of Σ, we present an evolution equation approach for system Ω only within the dual state concept. This dual state concept is obtained via the transformation of Ω, (47) into the equation

$$\tilde{\Omega} \qquad \boxed{\begin{aligned} \dot{x}(t) &= \int_{-t}^{0} d\eta(\tau)x(t+\tau) + \int_{-t}^{0} d\mu(\tau)\dot{x}(t+\tau) \\ &\quad + \int_{-t}^{0} d\beta(\tau)u(t+\tau) + f^1(-t), \\ x(0) &= f^0. \end{aligned}}$$

where $f = (f^0, f^1) \in M^p$ is given by

$$f^0 = \phi(0), \tag{48.1}$$

$$f^1(\sigma) = \int_{-h}^{\sigma} d\eta(\tau)\phi(\tau-\sigma) + \int_{-h}^{\sigma} d\mu(\tau)\dot{\phi}(\tau-\sigma) + \int_{-h}^{\sigma} d\beta(\tau)\xi(\tau-\sigma) \tag{48.2}$$

for $-h \leq \sigma \leq 0$. The same arguments as in Remark 2.1.4 show that $\tilde{\Omega}$ admits a unique solution $x(t)$, $t \geq 0$, for every $f \in M^p$ and every input $u \in L^p_{loc}([0,\infty);\mathbb{R}^m)$. The pair $f \in M^p$ is regarded as the initial state of $\tilde{\Omega}$. The corresponding state $(x(t), x^t) \in M^p$ at time $t \geq 0$ is given by

$$\begin{aligned} x^t(\sigma) &= \int_{\sigma-t}^{\sigma} d\eta(\tau)x(t+\tau-\sigma) + \int_{\sigma-t}^{\sigma} d\mu(\tau)\dot{x}(t+\tau-\sigma) \\ &= \int_{\sigma-t}^{\sigma} d\beta(\tau)u(t+\tau-\sigma) + f^1(\sigma-t), \ -h \leq \sigma \leq 0. \end{aligned} \tag{49}$$

As we already know, this expression can be obtained by a time shift in $\tilde{\Omega}$.

The following result shows that, in the case $\Gamma = 0$, system $\tilde{\Omega}$ represents the restriction of $\tilde{\Sigma}$ to the dense subspace ran ι^{T*} of the state space $W^{-1,p}$. The proof is omitted since it is strictly analogous to that of Lemma 2.1.5.

2.3.2 <u>LEMMA</u>. *Let $\Gamma = 0$ and let $f \in M^p$, $\mathrm{f} \in M^p$, and $u \in L^p_{loc}([0,\infty);\mathbb{R}^m)$ be given. Moreover let $x(t)$, $t \geq 0$, be the unique solution of $\tilde{\Omega}$ and $w(t)$, $\mathrm{x}(t)$ the unique solution pair of $\tilde{\Sigma}$. Then $\mathrm{x}(t) = x(t)$ for every $t \geq 0$ if and only if $\pi \mathrm{f} = \iota^{T*} f$.*

The desired evolution equations for the systems $\tilde{\Sigma}$ and $\tilde{\Omega}$ can be obtained by means of the duality relation between $\tilde{\Sigma}$ and Ω^T respectively between $\tilde{\Omega}$ and Σ^T.

THE TRANSPOSED EQUATION

Transposition of matrices leads to an observed NFDE where the output is obtained via the bounded linear functionals B^T and Γ^T on C with values in $\mathbb{R}^m$. These are given by

$$B^T\psi = \int_{-h}^{0} d\beta^T(\tau)\psi(\tau), \ \Gamma^T\psi = \int_{-h}^{0} d\gamma^T(\tau)\psi(\tau), \ \psi \in C.$$

If $\Gamma = 0$, then the output depends on the derivative of the solution. This means that we have to work in the state space $W^{1,q}$, i.e., with the system

$$\Omega^T \qquad \boxed{\begin{aligned} \dot{x}(t) &= L^T x_t + M^T \dot{x}_t \\ y(t) &= B^T x_t + \Gamma^T \dot{x}_T \end{aligned}} \quad .$$

This system admits a unique solution for every initial function $x_0 = \psi \in W^{1,q}$. The output $y(t)$ of Ω^T makes sense as an element of $L^q_{loc}([0,\infty); \mathbb{R}^m)$ and depends continuously on the initial state $\psi \in W^{1,q}$ (compare Remark 1.2.1 (ii)).

If we want to extend the transposed system to the product space M^q, we have to ensure that the output does not depend on the derivative of the solution, i.e. $\Gamma = 0$. In this case the desired extension is represented by the system

$$\Sigma^T \qquad \boxed{\begin{aligned} \dot{z}(t) &= L^T x_t \\ x(t) &= z(t) + M^T x_t \\ y(t) &= B^T x_t \end{aligned}} \quad .$$

This system admits a unique solution for every initial condition (8) where $\psi \in M^q$. The corresponding output $y(t)$ makes sense as an element of $L^q_{loc}([0,\infty); \mathbb{R}^m)$ and depends continuously on the initial state $\psi \in M^q$.

3.4 <u>REMARK</u> Let $\Gamma = 0$ and $\psi \in W^{1,q}$. Then the output of Σ^T which corresponds to the initial state $\iota\psi \in M^q$ coincides with the output of Ω^T.

Note that the dual state concept for the systems Ω^T and Σ^T would lead to an additional function-component in the state of the system which is due to the delays in the output variables. This corresponds to the fact that the input has to be included in the state of the systems Σ and Ω (within the original state concept). In order to avoid this further complication, we restrict our study of the transposed equation to the original state concept.

3.4 REMARK The output of the systems Ω^T and Σ^T may be described via the linear map $\mathcal{B}^T$: dom $A^T \to \mathbb{R}^m$ which we define by

$$\mathcal{B}^T\psi = B^T\psi + \Gamma^T\dot{\psi}, \qquad \psi \in \text{dom } A^T \subset W^{1,q}.$$

Three special cases of the output operators B^T and Γ^T are of particular importance for the properties of this map.

(i) If Γ^T consists only of an integral term, i.e.

$$\Gamma^T\psi = \int_{-h}^{0} B_2^T(\tau)\psi(\tau)d\tau, \quad \psi \in C \tag{50}$$

where $B_2(\cdot) \in L^p([-h,0]; \mathbb{R}^{n\times m})$, then $\mathcal{B}^T$ can be extended to a bounded, linear operator from $W^{1,q}$ into $\mathbb{R}^m$.

(ii) In the case $\Gamma = 0$ the operator $\mathcal{B}^T : W^{1,q} \to \mathbb{R}^m$ satisfies the hypothesis (H3) of Chapter 1.3. This means that, for every $T > 0$, there exists some $b_T > 0$ such that the following inequality holds for every $\psi \in W^{1,q}$

$$||\mathcal{B}^T S^T(\cdot)\psi||_{q,T} \leq b_T \; ||\iota^T\psi||_{M^q}. \tag{51}$$

This follows from the fact that the output of Σ^T depends continuously on the initial state.

(iii) Let $B^T : C \to \mathbb{R}^m$ be given by

$$B^T\psi = B_0^T\left(\psi(0) - M^T\psi\right) + \int_{-h}^{0} B_1^T(\tau)\psi(\tau)d\tau, \; \psi \in C, \tag{52}$$

where $B_0 \in \mathbb{R}^{n\times m}$ and $B_1(\cdot) \in L^p([-h,0]; \mathbb{R}^{n\times m})$. Then the output operator $\mathcal{B}^T$ can be extended to a bounded, linear map on M^q which maps $\psi \in M^q$ into $B_0^T\psi^0 + \int_{-h}^{0} B_1^T(\tau)\psi^1(\tau)d\tau \in \mathbb{R}^m$.

DUALITY

The following duality theorem is the central result of Chapter 2. In particular the theory of the structural operators depends essentially on this result. Moreover, the evolution equations for $\tilde{\Sigma}$ and $\tilde{\Omega}$ will come out as an immediate consequence.

2.3.5 THEOREM. *Let* $u(\cdot) \in L^p_{loc}([0,\infty);\mathbb{R}^m)$ *be given.*

(i) *Let* $f \in M^p$ *and* $\psi \in W^{1,q}$. *Moreover, let* $\pi(w(t),w^t,x^t) \in W^{-1,p}$ *be the state of* $\tilde{\Sigma}$ *- defined by* (46) *- and let* $x(t)$ *be the unique solution of* Ω^T, (11) *with output* $y(t)$. *Then*

$$\langle\psi,\pi(w(t),w^t,x^t)\rangle = \langle x_t,\pi f\rangle + \int_0^t y^T(t-s)u(s)ds, \; t \geq 0.$$

(ii) *Let* $f \in M^p$ *and* $\psi \in M^q$. *Moreover, let* $(x(t),x^t) \in M^p$ *be the state of* $\tilde{\Omega}$ *- defined by* (49) *- and let* $z(t)$, $x(t)$ *be the unique solution of* Σ^T, (8) *with output* $y(t)$. *Then*

$$\langle\psi,(x(t),x^t)\rangle = \langle(z(t),x_t),f\rangle + \int_0^t y^T(t-s)u(s)ds, \; t \geq 0.$$

<u>Proof</u>. (i) Let $x(t) = 0$ and $u(t) = 0$ for $t < 0$. Then

$$\begin{aligned}
&\int_0^t \left(x^T(t-s)Lx_s - [L^T x_{t-s}]^T x(s)\right)ds \\
&= \int_{-h}^0 \int_0^t x^T(t-s)d\eta(\tau)x(s+\tau)ds - \int_{-h}^0 \int_0^t x^T(t-s+\tau)d\eta(\tau)x(s)ds \\
&= -\int_{-h}^0 \int_{t+\tau}^t x^T(t+\tau-s)d\eta(\tau)x(s)ds \qquad (53) \\
&= -\int_{-h}^0 \int_\tau^0 \psi^T(\tau-\sigma)d\eta(\tau)x(t+\sigma)d\sigma.
\end{aligned}$$

Analogous expressions hold for M, B, and Γ. Moreover,

$$\begin{aligned}
\psi^T(0)w(t) - x^T(t)f^0 &= \int_0^t \frac{d}{ds} x^T(t-s)w(s)ds \\
&= \int_0^t x^T(t-s)\dot{w}(s)ds - \int_0^t \dot{x}^T(t-s)w(s)ds.
\end{aligned}$$

This implies

$$\langle\psi,\pi(w(t),w^t,x^t)\rangle$$

$$= \int_{-h}^{0} \psi^T(\sigma)w^t(\sigma)d\sigma + \int_{-h}^{0} \dot{\psi}^T(\sigma)x^t(\sigma)d\sigma + \psi^T(0)w(t)$$

$$= \int_{-h}^{0}\int_{\tau}^{0} \psi^T(\tau-\sigma)d\eta(\tau)x(t+\sigma)d\sigma + \int_{-h}^{0}\int_{\tau}^{0} \psi^T(\tau-\sigma)d\beta(\tau)u(t+\sigma)d\sigma$$

$$+ \int_{-h}^{0}\int_{\tau}^{0} \dot{\psi}^T(\tau-\sigma)d\mu(\tau)x(t+\sigma)d\sigma + \int_{-h}^{0}\int_{\tau}^{0} \dot{\psi}^T(\tau-\sigma)d\gamma(\tau)u(t+\sigma)d\sigma$$

$$+ \int_{-h}^{0} \psi^T(\sigma)f^1(\sigma-t)d\sigma + \int_{-h}^{0} \dot{\psi}^T(\sigma)f^2(\sigma-t)d\sigma + x^T(t)f^0$$

$$+ \int_{0}^{t} x^T(t-s)\Big(Lx_s + Bu_s + f^1(-s)\Big)ds$$

$$- \int_{0}^{t} \dot{x}^T(t-s)\Big(x(s) - Mx_s - \Gamma u_s - f^2(-s)\Big)ds$$

$$= x^T(t)f^0 + \int_{-t}^{0} x^T(t+\tau)f^1(\tau)d\tau + \int_{-t}^{0} \dot{x}^T(t+\tau)f^2(\tau)d\tau$$

$$+ \int_{-h}^{-t} \psi^T(t+\tau)f^1(\tau)d\tau + \int_{-h}^{-t} \dot{\psi}^T(t+\tau)f^2(\tau)d\tau$$

$$+ \int_{0}^{t} [L^T x_{t-s}]^T x(s)ds + \int_{0}^{t} [M^T\dot{x}_{t-s}]^T x(s)ds - \int_{0}^{t} \dot{x}^T(t-s)x(s)ds$$

$$+ \int_{0}^{t} [B^T x_{t-s}]^T u(s)ds + \int_{0}^{t} [\Gamma^T\dot{x}_{t-s}]^T u(s)ds$$

$$= \langle x_t,\pi f\rangle + \int_{0}^{t} y^T(t-s)u(s)ds.$$

(ii) Let us define $x(t) = 0$, $\dot{x}(t) = 0$, and $u(t) = 0$ for $t < 0$. Moreover, note that

$$\psi^{0^T}x(t) - z^T(t)f^0 = \int_0^t \frac{d}{ds} z^T(t-s)x(s)ds$$

$$= \int_0^t z^T(t-s)\dot{x}(s)ds - \int_0^t \dot{z}^T(t-s)x(s)ds.$$

Hence we obtain (by the use of equations analogous to (53))

$$\langle\psi,(x(t),x^t)\rangle$$

$$= \int_{-h}^0 \psi^{1^T}(\sigma)x^t(\sigma)d\sigma + \psi^{0^T}x(t)$$

$$= \int_{-h}^0 \int_\tau^0 \psi^{1^T}(\tau-\sigma)d\eta(\tau)x(t+\sigma)d\sigma + \int_{-h}^0 \int_\tau^0 \psi^{1^T}(\tau-\sigma)d\mu(\tau)\dot{x}(t+\sigma)d\sigma$$

$$+ \int_{-h}^0 \int_\tau^0 \psi^{1^T}(\tau-\sigma)d\beta(\tau)u(t+\sigma)d\sigma + \int_{-h}^0 \psi^{1^T}(\sigma)f^1(\sigma-t)d\sigma$$

$$+ z^T(t)f^0 + \int_0^t \left(x(t-s) - M^Tx_{t-s}\right)^T\dot{x}(s)ds$$

$$- \int_0^t [L^Tx_{t-s}]^Tx(s)ds$$

$$= z^T(t)f^0 + \int_{-h}^{-t} \psi^{1^T}(t+\tau)f^1(\tau)d\tau + \int_0^t x^T(t-s)\dot{x}(s)ds$$

$$- \int_0^t x^T(t-s)Lx_s ds - \int_0^t x^T(t-s)M\dot{x}_s ds$$

$$- \int_0^t \left(x^T(t-s)Bu_s - [B^Tx_{t-s}]^Tu(s)\right)ds$$

$$= z^T(t)f^0 + \int_{-h}^{-t} \psi^{1^T}(t+\tau)f^1(\tau)d\tau + \int_0^t y^T(t-s)u(s)ds$$

$$+ \int_0^t x^T(t-s)\left(\dot{x}(s) - Lx_s - M\dot{x}_s - Bu_s\right)ds$$

$$= \langle(z(t),x_t),f\rangle + \int_0^t y^T(t-s)u(s)ds. \qquad \square$$

For NFDEs in the product space framework, a duality result in the form of the previous theorem has not yet been developed in the literature. Related results for systems with undelayed input and output variables may be found in the references cited at the end of Section 2.1. For RFDEs with input delays we refer to Vinter-Kwong [147] and Delfour [28].

As a consequence of Theorem 2.3.5 we obtain the following infinite-dimensional variation-of-constants formulae for the systems $\tilde{\Sigma}$ and $\tilde{\Omega}$. For retarded systems such a result has been stated without proof in Delfour [28, Theorem 3.2].

2.3.6 THEOREM. *Let* $u \in L^p_{loc}([0,\infty);\mathbb{R}^m)$ *be given.*

(i) *Let* Γ *be of the form* (50) *and* $f \in M^p$. *Then the corresponding state of* $\tilde{\Sigma}$ *at time* $t \geq 0$ *is given by*

$$\pi(w(t),w^t,x^t) = S^{T*}(t)\pi f + \int_0^t S^{T*}(t-s)B^{T*}u(s)ds.$$

(ii) *Let* $\Gamma = 0$ *and* $f \in M^p$. *Then the corresponding state of* $\tilde{\Omega}$ *at time* $t \geq 0$ *is given by*

$$(x(t),x^t) = S^{T*}(t)f + \iota^{T*-1}\int_0^t S^{T*}(t-s)B^{T*}\,u(s)ds.$$

Proof. (i) Let $\psi \in W^{1,q}$. Then the corresponding output of Ω^T is given by $y(t) = B^T S^T(t)\psi$ for $t \geq 0$ (Remark 2.3.4). Hence it follows from Theorem 2.3.5 that

$$\langle\psi,\pi(w(t),w^t,x^t)\rangle_{W^{1,q},W^{-1,p}}$$

$$= \langle S^T(t)\psi,\pi f\rangle_{W^{1,q},W^{-1,p}} + \int_0^t \langle B^T S^T(t-s)\psi,u(s)\rangle_{\mathbb{R}^m}\,ds$$

$$= \langle\psi,S^{T*}(t)\pi f + \int_0^t S^{T*}(t-s)B^{T*}\,u(s)ds\rangle_{W^{1,q},W^{-1,p}}.$$

(ii) Let $\psi \in W^{1,q}$ and $\Gamma = 0$. Then the output of Σ^T corresponding to the initial state $\iota^T\psi \in M^q$ is given by $y(t) = B^T S^T(t)\psi$ for $t \geq 0$ (Remark 2.3.3). Hence it follows from Theorem 2.3.5 that

$$\langle \psi, \iota^{T*}(x(t),x^t\rangle_{W^{1,q},W^{-1,p}} = \langle \iota^T\psi,(x(t),x^t)\rangle_{M^q,M^p}$$

$$= \langle S^T(t)\iota^T\psi, f\rangle_{M^q,M^p} + \int_0^t \langle \mathcal{B}^T S^T(t-s)\psi, u(s)\rangle_{\mathbb{R}^m} \, ds$$

$$= \langle \psi, \iota^{T*} S^{T*}(t)f + \int_0^t S^{T*}(t-s)\mathcal{B}^{T*}u(s)\, ds\rangle_{W^{1,q},W^{-1,p}}.$$

Statement (ii) follows also from (i) and Lemma 2.3.2. □

The above theorem shows that $\tilde{\Sigma}$ and $\tilde{\Omega}$ are related to the Cauchy problem

$$d/dt\ \hat{x}(t) = A^{T*}\hat{x}(t) + \mathcal{B}^{T*}\ u(t) \tag{54}$$

(in the Banach spaces $W^{-1,p}$ and M^p) in the following way.

SYSTEM $\tilde{\Sigma}$ Let Γ be given by (50). Then, by Remark 2.3.4 (i), $\mathcal{B}^{T*}$ is a bounded, linear operator on $\mathbb{R}^m$ with values in $W^{-1,p}$. Theorem 2.3.6 (i) shows that in this case the state $\hat{x}(t) = \pi(w(t),w^t,x^t) \in W^{-1,p}$ of $\tilde{\Sigma}$, defined by (46), is a mild solution of (54).

2.3.7 REMARK Let $\Gamma = 0$. Then it follows from Remark 2.3.4 (ii) and Lemma 1.3.5 that the input operator $\mathcal{B}^{T*} : \mathbb{R}^m \to W^{-1,p}$ satisfies hypothesis (H2) of Section 1.3. This means that

$$\int_0^T S^{T*}(t-s)\mathcal{B}^{T*}u(s)ds \in \operatorname{ran} \iota^{T*}$$

and

$$\| \iota^{T*-1} \int_0^t S^{T*}\ (t-s)\mathcal{B}^{T*}\ u(s)ds \|_{M^p} \leqslant b_T \|u\|_{p,T} \tag{55}$$

for every $T > 0$ and every $u \in L^p([0,T];\mathbb{R}^m)$.

SYSTEM $\tilde{\Omega}$ Let $\Gamma = 0$ and let $\hat{x}(t) = (x(t),x^t) \in M^p$ be the state of $\tilde{\Omega}$ at time $t \geqslant 0$. Then it follows from Theorem 2.3.6 (ii) and Theorem 1.3.4 that $\hat{x}(t)$ is a solution of (54) in the sense of Definition 1.3.3. This means that $\hat{x}(t)$ is continuous in M^p and that $\hat{x}(t) = \iota^{T*}\hat{x}(t)$ is absolutely continuous in $W^{-1,p}$ and satisfies (54) for almost every $t \geqslant 0$.

So far we have described the systems Σ and Ω only within the dual state concept and in the case that Γ is given by (50). A description of Σ and Ω within the original state concept and in the general case can be given through the structural operators. These allow also a representation of the output of the transposed systems Ω^T and Σ^T.

THE STRUCTURAL OPERATORS

The equations (45) and (48) suggest the introduction of the structural operators $E : L^p \to W^{-1,p}$ and $\mathcal{E} : L^p \to M^p$. These do precisely the same job as the operators F and $\mathcal{F}$, namely they replace the initial function $\xi \in L^p$ of the input by the corresponding inhomogeneous term of system $\tilde{\Omega}$ (the operator $\mathcal{E}$) respectively by the bounded linear functional on $W^{1,q}$ which is represented by the forcing term of system $\tilde{\Sigma}$ (the operator E). Given $\xi \in L^p$, we define

$$E\xi = \pi f \in W^{-1,p}, \ f^0 = 0,$$

$$f^1(\sigma) = \int_{-h}^{\sigma} d\beta(\tau)\xi(\tau-\sigma), \ f^2(\sigma) = \int_{-h}^{\sigma} d\gamma(\tau)\xi(\tau-\sigma),$$

and

$$[\mathcal{E}\xi]^0 = 0, \ [\mathcal{E}\xi]^1(\sigma) = \int_{-h}^{\sigma} d\beta(\tau)\xi(\tau-\sigma),$$

for $-h \leqslant \sigma \leqslant 0$.

Operators of this type have been introduced for retarded systems by Vinter-Kwong [147] and by Delfour [28]. In particular, the following result has been proved in Vinter [147, Theorem 2.5.1] for retarded systems of the form $\dot{x}(t) = Lx_t + Bu_t$ where B is given by (52).

2.3.8 <u>COROLLARY</u>. *Let* $u \in L^p_{loc}([0,\infty);\mathbb{R}^m)$ *and* $\xi \in L^p$ *be given.*

(i) *Let* Γ *be given by* (50). *If* $\phi \in M^p$ *and the pair* w(t), x(t) *is the unique solution of* Σ, (42), *then*

$$F(w(t),x_t) + Eu_t = S^{T*}(t)[F\phi + E\xi] + \int_0^t \mathcal{S}^{T*}(t-s)\mathcal{B}^{T*}u(s)ds.$$

(ii) *Let* $\Gamma = 0$. *If* $\phi \in W^{1,p}$ *and* $x(t)$, $t \geq -h$, *is the unique solution of* Ω, (47), *then*

$$\mathcal{F}x_t + \mathcal{E}u_t = \mathcal{S}^{T*}(t)[\mathcal{F}\phi + \mathcal{E}\xi] + \iota^{T*-1} \int_0^t \mathcal{S}^{T*}(t-s)\mathcal{B}^{T*}u(s)ds.$$

In order to give an explicit description of the solutions to system Σ, respectively Ω, we have to introduce (finally) another structural operator $D : L^p \to W^{-1,p}$ respectively $\mathcal{D} : L^p \to M^p$. This operator describes the action of the input segment u_h on the forcing term of the respective equation. Given $\xi \in L^p$, we define

$$D\xi = \pi f \in W^{-1,p},\ f^0 = 0,$$

$$f^1(\sigma) = \int_\sigma^0 d\beta(\tau)\xi(\tau-\sigma-h),\ f^2(\sigma) = \int_\sigma^0 d\gamma(\tau)\xi(\tau-\sigma-h),$$

and

$$[\mathcal{D}\xi]^0 = 0,\ [\mathcal{D}\xi]^1(\sigma) = \int_\sigma^0 d\beta(\tau)\xi(\tau-\sigma-h),$$

for $-h \leq \sigma \leq 0$ (compare the right-hand side of the equations $\tilde{\Sigma}$ and $\tilde{\Omega}$).

The following result is an immediate consequence of the definition of the structural operators.

2.3.9 PROPOSITION. *Let* $u \in L^p([0,h];\mathbb{R}^m)$ *and* $\xi \in L^p$ *be given.*

(i) *Let* $\phi \in M^p$ *and let* $w(t)$, $0 \leq t \leq h$, *and* $x(t)$, $-h \leq t \leq h$, *be the corresponding solution of* Σ, (42), *then*

$$(w(h),x_h) = G[F\phi + E\xi + Du_h]. \tag{56}$$

(ii) *Let* $\phi \in W^{1,p}$ *and let* $x(t)$, $-h \leq t \leq h$, *be the unique solution of* Ω, (47), *then*

$$x_h = \mathcal{G}[\mathcal{F}\phi + \mathcal{E}\xi + \mathcal{D}u_h]. \tag{57}$$

The output of the systems Ω^T and Σ^T can be described by means of the operators

$$E^* : W^{1,q} \to L^q \qquad \mathcal{E}^* : M^q \to L^q$$

$$D^* : W^{1,q} \to L^q \qquad \mathcal{D}^* : M^q \to L^q$$

in the following way.

2.3.10 PROPOSITION

(i) *Let* $\psi \in W^{1,q}$. *Then the corresponding output* $y(t)$, $0 \leqslant t \leqslant h$, *of* Ω^T *is given by*

$$y(t) = [E^*\psi + D^*G^*F^*\psi](-t), \; 0 \leqslant t \leqslant h. \tag{58}$$

(ii) *Let* $\psi \in M^q$. *Then the corresponding output* $y(t)$, $0 \leqslant t \leqslant h$, *of* Ω^T *is given by*

$$y(t) = [\mathcal{E}^*\psi + \mathcal{D}^*\mathcal{G}^*\mathcal{F}^*\psi](-t), \; 0 \leqslant t \leqslant h. \tag{59}$$

PROOF. The following representation of the operators E^*, D^*, $\mathcal{E}^*$, $\mathcal{D}^*$ can be proved straightforwardly from the definition of the operators E, D, $\mathcal{E}$, $\mathcal{D}$. Given $\psi \in W^{1,q}$, we have

$$[E^*\psi](\sigma) = \int_{-h}^{\sigma} d\beta^T(\tau)\psi(\tau-\sigma) + \int_{-h}^{\sigma} d\gamma^T(\tau)\dot{\psi}(\tau-\sigma),$$

$$[D^*\psi](\sigma) = \int_{\sigma}^{0} d\beta^T(\tau)\psi(\tau-\sigma-h) + \int_{\sigma}^{0} d\gamma^T(\tau)\dot{\psi}(\tau-\sigma-h),$$

for $-h \leqslant \sigma \leqslant 0$. This proves (58). Given $\psi \in M^q$, we have

$$[\mathcal{E}^*\psi](\sigma) = \int_{-h}^{\sigma} d\beta^T(\tau)\psi^1(\tau-\sigma), \; [\mathcal{D}^*\psi](\sigma) = \int_{\sigma}^{0} d\beta^T(\tau)\psi(\tau-\sigma-h),$$

for $-h \leqslant \sigma \leqslant 0$. This proves (59). □

We close this section with a result on the operators E, D, $\mathcal{E}$, $\mathcal{D}$ which is analogous to Lemma 2.2.9.

2.3.11 LEMMA. *Let* $\Gamma = 0$. *Then*

$$E = \iota^{T*}\mathcal{E} \qquad E^* = \mathcal{E}^*\iota^T$$

$$D = \iota^{T*}\mathcal{D} \qquad D^* = \mathcal{D}^*\iota^T$$

Proof. If $\Gamma = 0$, then the following equations hold for every $\xi \in L^p$ and every $\psi \in W^{1,p}$

$$\langle\psi,E\xi\rangle = \langle\iota^T\psi,\mathcal{E}\xi\rangle = \int_{-h}^{0}\int_{\tau}^{0} \psi^T(\tau-\sigma)d\beta(\tau)\xi(\sigma)d\sigma,$$

$$\langle\psi,D\xi\rangle = \langle\iota^T\psi,\mathcal{D}\xi\rangle = \int_{-h}^{0}\int_{-h}^{\tau} \psi^T(\tau-\sigma-h)d\beta(\tau)\xi(\sigma)d\sigma. \qquad \square$$

2.4 SPECTRAL THEORY

Most of the results of this section are well known for RFDEs (Shimanov [140], Hale [42], Banks-Burns [1], Vinter [145], Delfour-Manitius [29]) and for NFDEs in the function spaces C (Hale-Meyer [43], Kappel [67]) and $W^{1,2}$ (Henry [48]). However, the proofs of these results which are available in the literature are rather complicated. It is the purpose of this section to develop a simple approach to the main facts in the spectral theory of NFDEs by means of the structural operators. In particular, we simplify some of the proofs in Delfour-Manitius [29] where an analogous theory was developed for RFDEs in the state space M^2.

Throughout this section all spaces and operators will be replaced by their obvious complex extensions. Correspondingly, the duality pairing between M^q and M^p $(1/p + 1/q = 1)$ is given by

$$\langle\psi,\phi\rangle = \psi^{0*}\phi^0 + \int_{-h}^{0} \psi^{1*}(\tau)\phi^1(\tau)d\tau$$

$(\phi \in M^p,\ \psi \in M^q)$ where $z^* = \bar{z}^T$ denotes the conjugate transposed of any complex vector (or matrix) z. By analogy, we extend the hereditary product $\langle\langle\ .,.\ \rangle\rangle$, and define

$$\langle\psi,\pi f\rangle = \psi^*(0)f^0 + \int_{-h}^{0} \psi^*(\tau)f^1(\tau)d\tau + \int_{-h}^{0} \dot{\psi}^*(\tau)f^2(\tau)d\tau$$

$(f \in \mathcal{M}^p,\ \psi \in W^{1,q})$ as well as

$$\langle\pi^T g,\phi\rangle = g^{0*}\phi(0) + \int_{-h}^{0} g^{1*}(\tau)\phi(\tau)d\tau + \int_{-h}^{0} g^{2*}(\tau)\dot{\phi}(\tau)d\tau$$

$(\phi \in W^{1,p},\ g \in \mathcal{M}^q)$.

Let us begin with a representation of the operators $\lambda I - A$ and $\lambda I - \mathcal{A}$ by means of the characteristic matrix of the NFDE (1) which is given by

$$\begin{aligned}\Delta(\lambda) &= \lambda[I - M(e^{\lambda\cdot})] - L(e^{\lambda\cdot}) \\ &= \lambda I - \int_{-h}^{0} e^{\lambda\tau} d\eta(\tau) - \lambda \int_{-h}^{0} e^{\lambda\tau} d\mu(\tau), \ \lambda \in \mathbb{C}.\end{aligned} \tag{60}$$

The proof of this basic result is straightforward and has been given by Henry [48] in the state space $W^{1,2}$ and recently by Ito [59, Theorem 2.2.8] in the state space M^2 (compare also Theorem 1.2.7 and Lemma 5.2.3).

2.4.1. LEMMA. *Let* $\lambda \in \mathbb{C}$ *be given.*

(i) *Let* $\phi, \Phi \in M^p$. *Then* $\phi \in$ dom A *and* $(\lambda I - A)\phi = \Phi$ *if and only if*

$$\phi^1(\tau) = e^{\lambda\tau}\phi^1(0) + \int_{\tau}^{0} e^{\lambda(\tau-\sigma)}\Phi^1(\sigma)d\sigma, \ -h \leqslant \tau \leqslant 0,$$

$$\phi^0 = \phi^1(0) - M\phi^1,$$

$$\Delta(\lambda)\phi^1(0) = \langle\langle e^{\bar{\lambda}\cdot}, F\Phi\rangle.$$

(ii) *Let* $\phi, \Phi \in W^{1,p}$. *Then* $\phi \in$ dom A *and* $(\lambda I - A)\phi = \Phi$ *if and only if*

$$\phi(\tau) = e^{\lambda\tau}\phi(0) + \int_{\tau}^{0} e^{\lambda(\tau-\sigma)} \Phi(\sigma)d\sigma, \ -h \leqslant \tau \leqslant 0,$$

$$\Delta(\lambda)\phi(0) = \langle\langle e^{\bar{\lambda}\cdot}, F\iota\Phi \rangle$$

Our next step is a concrete formula for the resolvent operators $(\lambda I - A)^{-1}$ and $(\lambda I - A)^{-1}$ analogous to Manitius [91, Proposition 2.1] and Delfour-Manitius [29, Theorem 4.4]. For this purpose we introduce the linear transformations

$$E_\lambda : \mathbb{C}^n \to W^{1,p}, \ H_\lambda : W^{-1,p} \to \mathbb{C}^n,$$

$$T_\lambda : M^p \to W^{1,p}$$

by defining

$$[E_\lambda x](\tau) = e^{\lambda\tau}x, \quad H_\lambda \pi f = \langle e^{\bar{\lambda}\cdot}, \pi f\rangle_{W^{1,q},W^{-1,p}},$$

$$[T_\lambda \phi](\tau) = \int_\tau^0 e^{\lambda(\tau-\sigma)}\phi^1(\sigma)d\sigma, \quad -h \leqslant \tau \leqslant 0, \tag{61}$$

for $x \in \mathbb{C}^n$, $f \in M^p$, $\phi \in M^p$. Then the theorem below is an immediate consequence of Lemma 2.4.1.

2.4.2 THEOREM.

(i) *The operators* A *and* A *have a pure point spectrum given by* $\sigma(\mathrm{A}) = \sigma(A) = \{\lambda \in \mathbb{C} | \det \Delta(\lambda) = 0\}$.

(ii) *Let* $\det \Delta(\lambda) \neq 0$. *Then the operators*

$$(\lambda I - \mathrm{A})^{-1} = \iota E_\lambda \Delta(\lambda)^{-1} H_\lambda F + \iota T_\lambda,$$

$$(\lambda I - A)^{-1} = E_\lambda \Delta(\lambda)^{-1} H_\lambda F \iota + T_\lambda \iota$$

are compact.

THE GENERALIZED EIGENSPACES

By Theorem 2.4.2, we can apply the general spectral theory of operators with a compact resolvent to our situation (see, e.g., Hille-Phillips [50, Section 5.14]). The bridge between the general functional analytic results and those on delay systems is given by the structural operators. In particular, we have the following relations between the generalized eigenspaces of A, A, A^{T*}, A^{T*}. For RFDEs in the product space M^2, results of this type have been proved in Delfour-Manitius [29] and Manitius [93].

2.4.3 LEMMA. *Let* $\lambda \in \sigma(\mathrm{A})$ *and* $k \in \mathbb{N}$. *Then*

$$\mathrm{F} \ker (\lambda I-\mathrm{A})^k = \ker (\lambda I-\mathrm{A}^{T*})^k, \quad \ker (\lambda I-\mathrm{A})^k = \mathrm{G} \ker (\lambda I-\mathrm{A}^{T*})^k.$$

$$F \ker (\lambda I-A)^k = \ker (\lambda I-A^{T*})^k, \quad \ker (\lambda I-A)^k = G \ker (\lambda I-A^{T*})^k.$$

$$\ker (\lambda I-\mathrm{A})^k = \iota \ker (\lambda I-A)^k, \quad \ker (\lambda I-A^{T*})^k = \iota^{T*} \ker (\lambda I-\mathrm{A}^{T*})^k.$$

Proof. First let $\phi \in \text{dom } A$. Then it follows from Theorem 2.2.2 and Lemma 1.3.8 that $F\phi \in \text{dom } A^{T*}$ and $A^{T*}F\phi = FA\phi$. By induction, we obtain for $k \in \mathbb{N}$

$$(\lambda I\text{-}A^{T*})^k F\phi = F(\lambda I\text{-}A)^k \phi, \ \phi \in \text{dom } A^k.$$

This shows that

$$F \ker (\lambda I\text{-}A)^k \subset \ker (\lambda I\text{-}A^{T*})^k.$$

The inclusion

$$G \ker (\lambda I\text{-}A^{T*})^k \subset \ker (\lambda I\text{-}A)^k$$

can be established by analogy.

Now we make use of the fact that the resolvent operator $(sI\text{-}A)^{-1}$ is compact. This implies that $\ker (\lambda I\text{-}A)^k$ is a finite-dimensional invariant subspace of the semigroup $S(t)$. Hence the operator $S(h) = GF$ is bijective on this subspace. We conclude that

$$\ker (\lambda I\text{-}A)^k = GF \ker (\lambda I\text{-}A)^k \subset G \ker (\lambda I\text{-}A^{T*})^k.$$

In the same manner we obtain, by the use of the equation $S^{T*}(h) = FG$, that

$$\ker (\lambda I\text{-}A^{T*})^k = FG \ker (\lambda I\text{-}A^{T*})^k \subset F \ker (\lambda I\text{-}A)^k.$$

This proves the first two assertions of the lemma. The second two can be established by analogy. The last two are trivial. □

2.4.4 COROLLARY. *Let* $\lambda \in \sigma(A^T)$ *and* $k \in \mathbb{N}$. *Then*

$$F^* \ker (\lambda I\text{-}A^T)^k = \ker (\lambda I\text{-}A^*)^k, \ \ker (\lambda I\text{-}A^T)^k = G^* \ker (\lambda I\text{-}A^*)^k.$$

$$F^* \ker (\lambda I\text{-}A^T)^k = \ker (\lambda I\text{-}A^*)^k, \ \ker (\lambda I\text{-}A^T)^k = G^* \ker (\lambda I\text{-}A^*)^k.$$

$$\ker (\lambda I\text{-}A^T)^k = \iota^T \ker (\lambda I\text{-}A^T)^k, \ \ker (\lambda I\text{-}A^*)^k = \iota^* \ker (\lambda I\text{-}A^*)^k.$$

For any $\lambda \in \sigma(A) = \sigma(A)$, let us now introduce the generalized eigenspaces

$$X_\lambda = \bigcup_{k \in \mathbb{N}} \ker (\lambda I\text{-}A)^k, \ X_\lambda = \bigcup_{k \in \mathbb{N}} (\lambda I\text{-}A)^k$$

of A and $\mathcal{A}$ as well as the complementary subspaces

$$X^{\lambda} = \bigcap_{k\in\mathbb{N}} \operatorname{ran}\,(\lambda I-A)^{k}, \quad \mathcal{X}^{\lambda} = \bigcap_{k\in\mathbb{N}} (\lambda I-\mathcal{A})^{k}.$$

In an analogous manner, $X_{\lambda}^{T}, X^{\lambda^{T}} \subset M^{q}$ and $\mathcal{X}_{\lambda}^{T}, \mathcal{X}^{\lambda^{T}} \subset W^{1,q}$ are associated with the operators A^{T} and $\mathcal{A}^{T}$. Some well-known properties of these subspaces are summarized below. They follow from the general theory of operators with a compact resolvent (see, e.g., Hille-Phillips [53, Theorem 5.14.3] and Taylor [142, Theorem 5.8A]).

2.4.5 <u>REMARKS</u>.

(i) For every $\lambda \in \sigma(A)$ there exists a minimal $k_{\lambda} \in \mathbb{N}$ such that

$$X_{\lambda} = \ker\,(\lambda I-A)^{k_{\lambda}}, \quad X^{\lambda} = \operatorname{ran}\,(\lambda I-A)^{k_{\lambda}},$$

$$\mathcal{X}_{\lambda} = \ker\,(\lambda I-\mathcal{A})^{k_{\lambda}}, \quad \mathcal{X}^{\lambda} = \operatorname{ran}\,(\lambda I-\mathcal{A})^{k_{\lambda}},$$

$$X_{\lambda}^{T} = \ker\,(\lambda I-A^{T})^{k_{\lambda}}, \quad X^{\lambda T} = \operatorname{ran}\,(\lambda I-A^{T})^{k_{\lambda}},$$

$$\mathcal{X}_{\lambda}^{T} = \ker\,(\lambda I-\mathcal{A}^{T})^{k_{\lambda}}, \quad \mathcal{X}^{\lambda^{T}} = \operatorname{ran}\,(\lambda I-\mathcal{A}^{T})^{k_{\lambda}},$$

Moreover, the subspaces on the left are finite-dimensional and those on the right are closed.

(ii) It follows from Lemma 2.4.3 that

$$\begin{aligned} \dim\ker\,(\lambda I-A)^{k} &= \dim\ker\,(\lambda I-A^{T*})^{k} \\ &= \dim\ker\,(\bar{\lambda} I-A^{T})^{k} \\ &= \dim\ker\,(\bar{\lambda} I-\mathcal{A}^{T})^{k} \end{aligned}$$

holds for all $\lambda \in \sigma(A)$ and $k \in \mathbb{N}$. Hence

$$\dim X_{\lambda} = \dim \mathcal{X}_{\lambda} = \dim X_{\bar{\lambda}}^{T} = \dim \mathcal{X}_{\bar{\lambda}}^{T}.$$

(iii)

$$M^p = X_\lambda \oplus X^\lambda,\ M^q = X_\lambda^T \oplus X^{\lambda^T},$$

$$W^{1,p} = X_\lambda \oplus X^\lambda,\ W^{1,q} = X^T \oplus X^{\lambda^T}.$$

(iv) The projection operator $P_\lambda : M^p \to X_\lambda$ associated with the above decomposition is given by

$$P_\lambda \phi = \frac{1}{2\pi i} \int_{\Gamma_\lambda} (sI-A)^{-1}\phi\, ds,\ \phi \in M^p,$$

where Γ_λ is a circle around λ, surrounding no other eigenvalue of A. The projection operators $P_\lambda : W^{1,p} \to X_\lambda$, $P_\lambda^T : M^q \to X_\lambda^T$, and $P_\lambda^T : W^{1,q} \to X_\lambda^T$ can be represented by analogy.

As a consequence of Corollary 2.2.4 we obtain a characterization of the complementary subspaces X^λ and X^λ via the generalized eigenspaces of the transposed equation.

2.4.6 THEOREM. *Let $\lambda \in \sigma(A)$ be given.*

(i) *Let $\phi \in M^p$. Then $\phi \in X^\lambda$ iff $F\phi \perp X_{\bar\lambda}^T$.*

(ii) *Let $f \in M^p$. Then $G\pi f \in X^\lambda$ iff $\pi f \perp X_{\bar\lambda}^T$.*

(iii) *Let $\phi \in W^{1,p}$. Then $\phi \in X^\lambda$ iff $F\phi \perp X_{\bar\lambda}^T$ or equivalently $\iota\phi \in X$.*

(iv) *Let $f \in M^p$. Then $Gf \in X^\lambda$ iff $f \perp X_{\bar\lambda}^T$.*

Proof. (i) It follows from Corollary 2.4.4 that $\phi \in X^\lambda = \text{ran}\,(\lambda I-A)^{k_\lambda}$ if and only if

$$\phi \perp \ker\,(\bar\lambda I-A^*)^{k_\lambda} = F^* \ker\,(\bar\lambda I-A^T)^{k_\lambda}.$$

Since $k_\lambda = k_{\bar\lambda}$, this is equivalent to

$$F\phi \perp \ker\,(\bar\lambda I-A^T)^{k_{\bar\lambda}} = X_{\bar\lambda}^T.$$

(ii) Let $f \in M^p$. Then $G\pi f \in X^\lambda$ if and only if $G\pi f$ annihilates $\ker\,(\bar\lambda I-A^*)^{k_\lambda}$ or equivalently

$$\pi f \perp G^* \ker (\bar{\lambda}I-A^*)^{k_{\bar{\lambda}}} = \ker (\bar{\lambda}I-A^T)^{k_{\bar{\lambda}}} = X_{\bar{\lambda}}^T$$

(Corollary 2.4.4).

(iii) Let $\phi \in W^{1,p}$. Then $\iota\phi \in X^\lambda$ if and only if $\iota\phi$ annihilates $\ker (\bar{\lambda}I-A^*)^{k_\lambda}$ or equivalently

$$\phi \perp \iota^* \ker (\bar{\lambda}I-A^*)^{k_\lambda} = \ker (\bar{\lambda}I-A)^{k_\lambda}$$

(Corollary 2.4.4). This means that

$$\phi \in \operatorname{ran} (\lambda I-A)^{k_\lambda} = X^\lambda.$$

The remainder of (iii) and (iv) can be proved in the same manner as (i) and (ii). □

The statements of Theorem 2.4.6 can also be formulated in terms of the hereditary product $\langle\langle .,. \rangle\rangle$ (Remark 2.2.10 (ii) and (iii)). This has been done by Shimanov [140], Hale [42], Banks-Burns [1], Hale-Meyer [43] and Henry [48]. In these papers the corresponding result is the difficult part of the theory. The properties of the spectral projection can then be proved in a simple straightforward way. This has been worked out for RFDEs in the state spaces C (Hale [42]) and M^2 (Banks-Burns [1], Delfour-Manitius [29]) and for NFDEs in the state spaces C (Hale-Meyer [43]) and $W^{1,2}$ (Henry [48]). Precisely the same arguments apply to NFDEs in the product space framework. Therefore we content ourselves with a summary of the main facts.

THE SPECTRAL PROJECTION

Let $\lambda \in \sigma(A)$ be given and let $N = \dim X_\lambda = \dim X_{\bar{\lambda}}^T$ (Remark 2.4.5 (ii)). Moreover let $\{\phi_1,\ldots,\phi_N\}$ be a basis of X_λ, $\{\psi_1,\ldots,\psi_N\}$ a basis of $X_{\bar{\lambda}}^T$, and introduce the matrix functions

$$\Phi = \begin{bmatrix} \phi_1 & \ldots & \phi_N \end{bmatrix} \in W^{1,p}([-h,0];\mathbb{C}^{n\times N}),$$

$$\Psi = \begin{bmatrix} \psi_1 & \ldots & \psi_N \end{bmatrix} \in W^{1,q}([-h,0];\mathbb{C}^{n\times N}).$$

Then the complex N N-matrix $\langle\Psi,F\iota\Phi\rangle = \langle\iota^T\Psi,F\Phi\rangle$ is nonsingular. Hence we can assume without loss of generality that

$$\langle\Psi,F\iota\Phi\rangle = \langle\iota^T\Psi,F\Phi\rangle = I. \tag{62}$$

As a direct consequence of this equation and Theorem 2.4.6 we obtain a representation of the spectral projections which correspond to the systems Σ and Ω, namely

$$\begin{aligned} P_\lambda\phi &= \iota\Phi\ \langle\Psi,F\phi\rangle,\ \phi \in M^p, \\ P_\lambda\phi &= \Phi\langle\iota^T\Psi,F\phi\rangle,\ \phi \in W^{1,p}. \end{aligned} \tag{63}$$

By analogy, the spectral projections associated with the transposed systems are given by

$$\begin{aligned} P^T_{\bar\lambda} &= \iota^T\Psi\ \langle F^*\psi,\Phi\rangle^*,\ \psi \in M^q, \\ P^T_{\bar\lambda} &= \Psi\langle F^*\psi,\iota\Phi\rangle^*,\ \ \psi \in W^{1,q}. \end{aligned} \tag{64}$$

We now study the projection of the systems Σ, Ω, Ω^T, Σ^T to their respective eigensubspaces. To do this we introduce the complex $N\times N$-matrix A_λ by

$$A\Phi = \Phi A_\lambda. \tag{65}$$

This matrix describes the dynamics of the spectral projection of the homogeneous systems Σ and Ω. More precisely, A_λ has the following well-known properties (Hale [42], Hale-Meyer [43], Henry [48]).

2.4.7 <u>PROPOSITION</u>. *Let* (62) *be satisfied and let* $A_\lambda \in \mathbb{C}^{N\times N}$ *be given by* (65). *Then*

(i) $A^T\Psi = \Psi A^*_\lambda$,

(ii) $\Phi(\tau) = \Phi(0)e^{A_\lambda\tau}$, $\Psi(\tau) = \Psi(0)e^{A^*_\lambda\tau}$, $-h \leqslant \tau \leqslant 0$,

(iii) $S(t)\Phi = \Phi e^{A_\lambda t}$, $S^T(t)\Psi = \Psi e^{A^*_\lambda t}$, $t \geqslant 0$,

(iv) $\sigma(A_\lambda) = \{\lambda\}$.

The action of the input on the spectral projection of the systems Σ and Ω is described by the complex $N\times m$-matrix

$$B_\lambda = \int_{-h}^{0} \Psi^*(\tau)d\beta(\tau) + A_\lambda\int_{-h}^{0} \Psi^*(\tau)d\gamma(\tau). \qquad (66)$$

2.4.8 PROPOSITION. *Let* $u \in L^p_{loc}([0,\infty);\mathbb{C}^m)$ *be given.*

(i) *Let* $f \in M^p$ *and let* $\pi(w(t),w^t,x^t) \in W^{-1,p}$ *be the corresponding state of* $\tilde{\Sigma}$, *defined by* (46). *Then* $x_\lambda(t) = \langle\Psi,\pi(w(t),w^t,x^t)\rangle \in \mathbb{C}^N$ *satisfies the ordinary differential equation*

$$\Sigma_\lambda \qquad \boxed{\dot{x}_\lambda(t) = A_\lambda x_\lambda(t) + B_\lambda u(t)}$$

(ii) *Let* $\Gamma = 0$, $f \in M^p$, *and let* $(x(t),x^t) \in M^p$ *be the corresponding state of* $\tilde{\Omega}$, *defined by* (49). *Then* $x_\lambda(t) = \langle\iota^T\Psi,(x(t),x^t)\rangle \in \mathbb{C}^N$ *satisfies* Σ_λ.

Proof. (i) First note that

$$B^*_\lambda = \int_{-h}^{0} d\beta^T(\tau)\Psi(\tau) + \int_{-h}^{0} d\gamma^T(\tau)\dot{\Psi}(\tau) = B^T\Psi + \Gamma^T\dot{\Psi} = \mathcal{B}^T\Psi \qquad (67)$$

(Remark 2.3.4). Hence it follows from Theorem 2.3.5 that

$$\begin{aligned} x_\lambda(t) &= \langle\mathcal{S}^T(t)\Psi,\pi f\rangle + \int_0^t \langle\mathcal{B}^T\mathcal{S}^T(t-s)\Psi,u(s)\rangle_{\mathbb{C}^m}\, ds \\ &= \langle\Psi e^{A^*_\lambda t},\pi f\rangle + \int_0^t \langle\mathcal{B}^T\Psi e^{A^*_\lambda(t-s)},u(s)\rangle_{\mathbb{C}^m}\, ds \\ &= e^{A_\lambda t}\langle\Psi,\pi f\rangle + \int_0^t e^{A_\lambda(t-s)} B_\lambda u(s)ds. \end{aligned}$$

(ii) follows from the fact that $\iota^{T*} : M^p \to W^{-1,p}$ maps the state $(x(t),x^t) \in M^p$ of $\tilde{\Omega}$ into the corresponding state $\pi(w(t),w^t,x^t) \in W^{-1,p}$ of $\tilde{\Sigma}$ if $\Gamma = 0$ (see Lemma 2.3.2). □

Let us now discuss the spectral projections of the different systems.

SYSTEM Σ Let $w(t)$, $x(t)$ be a solution pair of Σ. Then it follows from Proposition 2.4.8 that the function

$$x_\lambda(t) = \langle\Psi,F(w(t),x_t) + Eu_t\rangle \in \mathbb{C}^N$$

satisfies Σ_λ. We mention without proof that the map

$$(\phi,\xi) \to (\iota\Phi,0) \langle\Psi,F\phi + E\xi\rangle, \ (\phi,\xi) \in M^p \times L^p,$$

is the spectral projection of the semigroup on $M^p \times L^p$ which corresponds to the free motions of Σ ($u(t) = 0$ for $t \geq 0$).

<u>SYSTEM Ω</u> Let $\Gamma = 0$ and let $x(t)$ be a solution of Ω. Then it follows from Proposition 2.4.8 that

$$x_\lambda(t) = \langle\iota^T\Psi, Fx_t + Eu_t\rangle \in \mathbb{C}^N$$

satisfies Σ_λ. We mention without proof that the map

$$(\phi,\xi) \to (\Phi,0) \langle\iota^T\Psi, F\phi + E\xi\rangle, \ (\phi,\xi) \in W^{1,p} \times L^p,$$

is the spectral projection of the semigroup on $W^{1,p} \times L^p$ which corresponds to the free motions of Ω ($u(t) = 0$ for $t \geq 0$).

<u>SYSTEM Ω^T</u> Let $\psi \in W^{1,q}$ be given. Then it follows from (64) and Proposition 2.4.7 that $P_\lambda^T S^T(t)\psi = \Psi x_\lambda(t) \in X_\lambda^T$ where $x_\lambda(t) \in \mathbb{C}^N$ is of the form

$$\begin{aligned} x_\lambda(t) &= \langle\iota^T S^T(t)\psi, F\Phi\rangle^* = \langle\iota^T\psi, S^{T*}(t)F\Phi\rangle^* \\ &= \langle\iota^T\psi, FS(t)\Phi\rangle^* = \langle\iota^T\psi, F\Phi e^{A_\lambda t}\rangle^* \\ &= e^{A_\lambda^* t}\langle\iota^T\psi, F\Phi\rangle^*. \end{aligned}$$

Moreover the corresponding output is given by

$$y_\lambda(t) = B^T P_\lambda^T S^T(t)\psi = B^T\Psi x_\lambda(t) = B_\lambda^* x_\lambda(t)$$

(see equation (67)). Hence the pair $x_\lambda(t)$, $y_\lambda(t)$ satisfies

$$\Sigma_\lambda^* \qquad \boxed{\dot{x}_\lambda(t) = A_\lambda^* x_\lambda(t), \ y_\lambda(t) = B_\lambda^* x_\lambda(t)} \quad .$$

<u>SYSTEM Σ^T</u> Let $\Gamma = 0$ and let $\psi \in M^q$ be given. Then it follows again from (64) and Proposition 2.4.7 that $P_\lambda^T S^T(t)\psi = \iota^T\Psi x_\lambda(t) \in X_\lambda^T$ where

$$x_\lambda(t) = \langle S^T(t)\psi, F\Phi\rangle^* = e^{A^*_\lambda t} \langle \psi, F\Phi\rangle^*.$$

The reduced output is given by

$$y_\lambda(t) = B^T[P^T_\lambda S^T(t)\psi]^1 = B^T\Psi x_\lambda(t) = B^*_\lambda x_\lambda(t).$$

Hence the pair $x_\lambda(t)$, $y_\lambda(t)$ is again described by system Σ^*_λ.

THE FREQUENCY DOMAIN

We close this section with some results on the Laplace transform. We make use of the abbreviation $\hat{x}(s) = \int_0^\infty e^{-st}x(t)\,dt$ for the Laplace transform of a function $x(t)$ on the positive real axis.

2.4.9 PROPOSITION. *Let* $u \in L^p_{loc}([0,\infty);\mathbb{R}^m)$ *be Laplace transformable.*

(i) *If* $f \in M^p$ *and* $w(t)$, $x(t)$ *satisfy* $\tilde{\Sigma}$, *then*

$$\hat{x}(s) = \Delta(s)^{-1}\, [\langle e^{\bar{s}\cdot}, \pi f\rangle + [B(e^{s\cdot}) + s\Gamma(e^{s\cdot})]\hat{u}(s)]. \tag{68}$$

(ii) *If* $f \in M^p$ *and* $x(t)$ *satisfies* $\tilde{\Omega}$, *then*

$$\hat{x}(s) = \Delta(s)^{-1}[\langle e^{\bar{s}\cdot}, f\rangle + B(e^{s\cdot})\hat{u}(s)]. \tag{69}$$

(iii) *If* $\psi \in W^{1,q}$ *and* $y(t)$, $t \geq 0$, *is the corresponding output of* Ω^T, (11), *then*

$$\hat{y}(s) = [B^T(e^{s\cdot}) + s\Gamma^T(e^{s\cdot})]\, \Delta^T(s)^{-1}\langle F^*\bar{\psi}, \iota e^{s\cdot}\rangle^T + B^T(e^{s\cdot} * \psi) + \Gamma^T(e^{s\cdot} * \dot{\psi} - e^{s\cdot}\psi(0)). \tag{70}$$

(iv) *If* $\psi \in M^q$ *and* $y(t)$, $t \geq 0$, *is the corresponding output of* Σ^T, (8), *then*

$$\hat{y}(s) = B^T(e^{s\cdot})\, \Delta^T(s)^{-1}\langle F^*\bar{\psi}, e^{s\cdot}\rangle^T + B^T(e^{s\cdot} * \psi^1). \tag{71}$$

Proof. (i) It follows from Proposition 2.2.6 and Remark 2.2.5 (v) that $x(t)$ and $w(t)$ are Laplace transformable. Defining $x(t) := 0$ and $u(t) := 0$ for $t < 0$, we obtain

$$s\,\hat{w}(s) - f^0 - \int_0^h e^{-st} f^1(-t)dt = \int_0^\infty e^{-st}[Lx_t + Bu_t]dt,$$

$$\hat{w}(s) = \hat{x}(s) - \int_0^h e^{-st} f^2(-t)dt - \int_0^\infty e^{-st}[Mx_t + \Gamma u_t]dt.$$

This implies

$$s\,\hat{x}(s) - \langle e^{\bar{s}\cdot}, \pi f\rangle$$

$$= s\int_0^\infty e^{-st}[Mx_t + \Gamma u_t]dt + \int_0^\infty e^{-st}[Lx_t + Bu_t]dt$$

$$= s\int_{-h}^0 d\mu(\tau) \int_0^\infty e^{-st}x(t+\tau)dt + \int_{-h}^0 d\eta(\tau)\int_0^\infty e^{-st}x(t+\tau)dt$$

$$+ s\int_{-h}^0 d\gamma(\tau)\int_0^\infty e^{-st}u(t+\tau)dt + \int_{-h}^0 d\beta(\tau)\int_0^\infty e^{-st}u(t+\tau)dt$$

$$= [sM(e^{s\cdot}) + L(e^{s\cdot})]\hat{x}(s) + [s\Gamma(e^{s\cdot}) + B(e^{s\cdot})]\hat{u}(s).$$

(ii) follows from (i) and Lemma 2.3.2.

(iii) Let $x(t)$, $t \geqslant -h$, be the unique solution of Ω^T, (11). Then it follows from (ii) that

$$\hat{x}(s) = \Delta^T(s)^{-1}\,\langle F^*\bar{\psi}, \iota e^{s\cdot}\rangle^T$$

Moreover $\hat{\dot{x}}(s) = s\hat{x}(s) - \psi(0)$. This implies

$$\hat{y}(s) = \int_{-h}^0 d\beta^T(\tau)\int_0^\infty e^{-st}x(t+\tau)dt + \int_{-h}^0 d\gamma^T(\tau)\int_0^\infty e^{-st}\dot{x}(t+\tau)dt$$

$$= B^T(e^{s\cdot})\hat{x}(s) + \int_{-h}^0 d\beta^T(\tau)\int_\tau^0 e^{s\sigma}x(\tau-\sigma)d\sigma$$

$$+ \Gamma^T(e^{s\cdot})\hat{\dot{x}}(s) + \int_{-h}^0 d\gamma^T(\tau)\int_\tau^0 e^{s\sigma}\dot{x}(\tau-\sigma)d\sigma$$

$$= [B^T(e^{s\cdot}) + s\Gamma^T(e^{s\cdot})]\hat{x}(s) + B^T(e^{s\cdot} * \psi)$$

$$+ \Gamma^T(e^{s\cdot} * \dot{\psi} - e^{s\cdot}\psi(0)).$$

(iv) If $\psi \in \operatorname{ran} \iota^T$, then statement (iv) is a consequence of (iii) and

Remark 2.3.3. In general, (iv) follows from continuous dependence and the fact that ran ι^T is dense in M^q. □

The following characterization of the subspace $X_\lambda^\perp$, X^λ, $X_\lambda^\perp$, X^λ is of particular importance in connection with the above result on the Laplace transform.

2.4.10 THEOREM. *Let* $\lambda \in \sigma(A)$ *be given.*

(i) *Let* $g \in M^q$. *Then* $\pi^T g \perp X_\lambda$ *if and only if the function* $\langle\pi^T g\, e^{s\cdot}\rangle\, \Delta(s)^{-1}$, $s \in \mathbb{C}$, *is holomorphic at* $s = \lambda$.

(ii) *Let* $g \in M^q$. *Then* $g \perp X_\lambda$ *if and only if the function* $\langle g, \iota e^{s\cdot}\rangle \Delta(s)^{-1}$, $s \in \mathbb{C}$, *is holomorphic at* $s = \lambda$.

(iii) *Let* $\phi \in M^p$. *Then* $\phi \in X^\lambda$ *if and only if the function* $\Delta(s)^{-1}\langle e^{\bar{s}\cdot}, F\phi\rangle$, $s \in \mathbb{C}$, *is holomorphic at* $s = \lambda$.

(iv) *Let* $\phi \in W^{1,p}$. *Then* $\phi \in X^\lambda$ *if and only if the function* $\Delta(s)^{-1}\langle\iota^T e^{\bar{s}\cdot}, F\phi\rangle$, $s \in \mathbb{C}$, *is holomorphic at* $s = \lambda$.

Proof. (i) First note that $(A-\lambda I)^k P_\lambda = 0$ for every $k \geqslant k_\lambda$. By the use of this fact, it is easy to see that the following equation holds for every $s \notin \sigma(A)$

$$(sI-A)^{-1}P_\lambda = \sum_{k=0}^{k_\lambda -1} (s-\lambda)^{-k-1}(A-\lambda I)^k P_\lambda. \tag{72}$$

(compare Kato [68, Section III.6.5]). Now let $\pi^T g \perp X_\lambda = \operatorname{ran} P_\lambda$. Then it follows from (72) that

$$\langle\pi^T g, (sI-A)^{-1}\phi\rangle = \langle\pi^T g, (sI-A)^{-1}(I-P_\lambda)\phi\rangle \tag{73}$$

for every $s \notin \sigma(A)$ and every $\phi \in W^{1,q}$. This function is holomorphic at $s = \lambda$, since $\lambda \notin \sigma(A|_{X^\lambda})$. Conversely, suppose that the function on the left-hand side of (73) is holomorphic at $s = \lambda$ for every $\phi \in W^{1,p}$. Then it follows from Remark 2.4.5 (iv) that $\pi^T g \perp X_\lambda$.

Applying Theorem 2.4.2 (ii), we obtain that $\langle\pi^T g, (sI-A)^{-1}\phi\rangle$ is holomorphic at $s = \lambda$ for every $\phi \in W^{1,p}$ if and only if the complex function $\langle\pi^T g, e^{s\cdot}\rangle\, \Delta(s)^{-1}\, \langle e^{\bar{s}\cdot}, F\iota\phi\rangle$ is holomorphic at $s = \lambda$ for every $\phi \in W^{1,p}$. Since ran ι is dense in M^p, we may replace $\iota\phi$ by any element of M^p. We choose the pair $(x,0) \in M^p$ where x is an arbitrary complex n-vector. This proves (i).

(ii) follows from (i) and the fact that $g \perp X_\lambda = \iota X_\lambda$ if and only if $\iota^* g \perp X_\lambda$.

(iii) Let $\phi \in M^p$. Then $\phi \in X^\lambda$ if and only if $F\phi \perp X^T_{\bar{\lambda}}$ (Theorem 2.4.6) or, equivalently, the complex function

$$\langle e^{s\cdot}, F\phi\rangle^* \, \Delta^T(s)^{-1}, \; s \in \mathbb{C},$$

is holomorphic at $s = \bar{\lambda}$ (see (i)). This means that the function

$$\Delta(\bar{s})^{-1} \langle e^{s\cdot}, F\phi\rangle, \; s \in \mathbb{C},$$

is holomorphic at $s = \bar{\lambda}$ which proves (iii).

(iv) follows from (iii) and the fact that $\phi \in X^\lambda$ if and only if $\iota\phi \in X^\lambda$ (Theorem 2.4.6 (iii)). □

The main idea in the proof of the above theorem is due to Delfour and Manitius [29, Lemma 5.2], who proved the corresponding result for retarded systems.

3 Completeness and small solutions

3.1 COMPLETENESS OF EIGENFUNCTIONS AND NONEXISTENCE OF NONZERO SMALL SOLUTIONS

Throughout this chapter we denote by Σ, Ω, Σ^T, Ω^T, the homogeneous systems of Section 2.1. These systems will be studied in Section 3.1 within the original state concept.

For any homogeneous delay equation there are two fundamental questions concerning the structural properties of the system.

1* Under what conditions is the whole state space spanned by the generalized eigenfunctions of the system? (completeness)

2* Under what conditions do there exist nonzero solutions which vanish after a finite time? (small solutions)

It turns out that there is a duality between these two properties. Roughly speaking, we shall see that a NFDE is complete if and only if the transposed equation has no nonzero small solution. This duality relation was first discovered by Manitius [93] for retarded systems. Let us begin with a discussion of the property of completeness.

COMPLETENESS

The problem of completeness of eigenfunctions has been studied by Levinson-McCalla [85] for scalar retarded systems. A fairly complete theory for RFDEs in the product space framework has been presented by Manitius [93] and Delfour-Manitius [29]. For neutral systems an analogous theory has not yet been developed. Such a development has been cited by Ito [59] as a difficult open problem. However, it turns out that - within the framework of our state space approach in Chapter 2 - this theory becomes quite easy. Some results on completeness of NFDEs in the state space $W^{1,2}$ may be found in O'Connor [109], Jakubczyk [62] and Bartosiewica [9].

The property of completeness is of some importance in the optimal control theory (Banks-Manitius [8]), for the finite-dimensional compensator design

(Schumacher [136] and for the controllability and observability properties of NFDEs (Chapter 4).

For convenience we introduce the closed subspaces

$$X_\sigma = \mathrm{cl}(\mathrm{span}\ X_\lambda | \lambda \in \sigma(A)\}) \subset M^p$$

$$X_\sigma = \mathrm{cl}(\mathrm{span}\ \{X_\lambda | \lambda \in \sigma(A)\}) \subset W^{1,p}$$

and by analogy, with an obvious meaning, $X_\sigma^T \subset M^q$ and $X_\sigma^T \subset W^{1,q}$. Note that these can be interpreted both as real and as complex subspaces.

3.1.1 DEFINITION. *System* Σ *(respectively* Ω*) is said to be complete if* $X_\sigma = M^p$ *(respectively* $X_\sigma = W^{1,p}$*).*

As an immediate consequence of this definition together with Theorem 2.4.10 we obtain the following completeness criterion.

3.1.2 COROLLARY.

(i) *System* Σ *is not complete if and only if there exists some nonzero* $g \in M^q$ *such that the complex function* $\langle g, \iota e^{s\cdot} \rangle\ \Delta(s)^{-1}$, $s \in \mathbb{C}$, *is entire.*

(ii) *System* Ω *is not complete if and only if there exists some* $g \in M^q$ *such that* $\pi^T g \neq 0$ *and the complex function* $\langle \pi^T g, e^{s\cdot} \rangle\ \Delta(s)^{-1}$, $s \in \mathbb{C}$, *is entire.*

Statement (i) in the above corollary is a generalization of the corresponding result on RFDEs in Delfour-Manitius [29, Corollary 5.4]. A rather complicated proof of statement (ii) can be found in O'Connor [109, Lemma 4.1].

Let us now introduce the concept of small solutions (Henry).

SMALL SOLUTIONS

3.1.3 DEFINITION. *A solution pair* $w(t)$, $x(t)$ *of* Σ *is said to be small if*

$$\lim_{t\to\infty} e^{\omega t}\ \|(w(t),x_t)\|_{M^p} = 0$$

for every $\omega \geq 0$. *A solution* $x(t)$ *of* Ω *is said to be small if the corresponding solution pair* $x(t) = x(t)$, $t \geq -h$, *and* $w(t) = x(t) - Mx_t$, $t \geq 0$, *of Σ is small.*

In other words, a small solution to a NFDE tends to zero more rapidly than any exponential. Note that the Laplace transform of such a function is an entire function. The important fact is that any small solution to any delay equation vanishes after a finite time. This was first proved in the 'classical' paper of Henry [46] for RFDEs in the state space C. For retarded systems, this implies the analogous result in the product space M^p since every solution will be in the state space C after the time $t = h$. Moreover it has been indicated by Henry [49, 50] that a corresponding statement holds for neutral systems. A very nice proof for NFDEs in the state space C has been presented by Kappel [67]. Precisely the same arguments apply to system Σ. This leads to the following result.

3.1.4 THEOREM. *Let* $\phi \in M^p$ *be given and let* $w(t)$, $x(t)$ *be the corresponding solution pair of system* Σ. *Then the following statements are equivalent.*

(i) *The pair* $w(t)$, $x(t)$ *is a small solution of* Σ.

(ii) *The function* $\Delta(s)^{-1} \langle e^{\bar{s}\cdot}, F\phi\rangle$, $s \in \mathbb{C}$, *is entire.*

(iii) *There exists a (minimal) time* $T_\phi \geq -h$ *such that* $x(t) = 0$ *for every* $t \geq T_\phi$.

If (iii) *is satisfied, then*

$$T_\phi \leq (n-1)h - \alpha \tag{1}$$

where α *is the exponential growth of* $\det \Delta(s)$, *i.e.*

$$\alpha = \limsup_{|s| \to \infty} |s|^{-1} \log |\det \Delta(s)| \geq 0.$$

3.1.5 REMARK. The implication (iii) $\Rightarrow$ (i) in the previous theorem is trivial. Moreover, it follows from Proposition 2.4.9 that (i) implies (ii). The hard part of the theorem is to prove that (ii) implies (iii) and to get the estimate (1). This can be done with exactly the same arguments that are given in Kappel [67, Theorem 3.1]. For precise verification one needs the formula

$$\int_0^\infty e^{-st}x(t-h)dt = e^{-sh}\,\Delta(s)^{-1}\Big(\phi^0 + s\int_{-h}^{0} e^{-s\tau}\phi^1(\tau)d\tau \tag{2}$$

$$- \int_{-h}^{0} [d\eta(\tau) + sd\mu(\tau)] \int_{-h}^{\tau} e^{s(\tau-\sigma)}\phi^1(\sigma)d\sigma\Big)$$

which follows easily from (2.68).

Let us now introduce the subspaces

$$X_0 = \bigcap_{\lambda\in\sigma(A)} X^\lambda \subset M^p,\quad X_0^T = \bigcap_{\lambda\in\sigma(A^T)} X^{\lambda^T} \subset M^q,$$

$$\mathcal{X}_0 = \bigcap_{\lambda\in\sigma(A)} \mathcal{X}^\lambda \subset W^{1,p},\quad \mathcal{X}_0^T = \bigcap_{\lambda\in\,\sigma(A^T)} \mathcal{X}^{\lambda^T} \subset W^{1,q}.$$

Then $\phi \in X_0$ if and only if the function $\Delta(s)^{-1}\ \langle e^{\bar{s}\cdot}, F\phi\rangle$, $s \in \mathbb{C}$, is entire (Theorem 2.4.10). An analogous characterization can be given for $\mathcal{X}_0$. Hence it follows from Theorem 3.1.4 that X_0 and $\mathcal{X}_0$ are precisely the subspaces of those initial states (of Σ and Ω) which lead to small solutions of the respective system.

3.1.6 <u>COROLLARY</u>. *There exists a (minimal) time* $T_0 \leqslant nh - \alpha$ *such that*

$$X_0 = \ker S(t) \qquad X_0^T = \ker S^T(t)$$

$$\mathcal{X}_0 = \ker \mathcal{S}(t) \qquad \mathcal{X}_0^T = \ker \mathcal{S}^T(t)$$

for every $t \geqslant T_0$.

This corollary is the starting-point for the relations between the spectral properties and the small solutions of neutral systems. For the derivation of these duality results we need the following interrelations between the subspaces with index $_0$ and those with index $_\sigma$ by means of the structural operators. These relations follow immediately from Theorem 2.4.6.

3.1.7 <u>COROLLARY</u>.

(i) *Let* $\phi \in M^p$. *Then* $\phi \in X_0$ *iff* $F\phi \perp \mathcal{X}_\sigma^T$.

(ii) *Let* $f \in \mathcal{M}^p$. *Then* $G\pi f \in \mathcal{X}_0$ *iff* $\pi f \perp X_\sigma^T$.

(iii) *Let* $\phi \in W^{1,p}$. *Then* $\phi \in X_0$ *iff* $F\phi \perp X_\sigma^T$ *or equivalently* $\iota\phi \in X_0$.

(iv) *Let* $f \in M^p$. *Then* $Gf \in X_0$ *iff* $f \perp X_\sigma^T$.

This result allows us to dualize Corollary 3.1.6. We obtain that the closed span of the generalized eigenspaces is precisely the closure of the range of the semigroup operator if t is large enough. This has first been proved by Henry [46] for retarded systems in the state space C. The corresponding result in the product space M^2 can be found in Manitius [93]. For neutral systems in the state space C we refer to Henry [49].

3.1.8 PROPOSITION. *For every* $t \geqslant T_0$

$$X_\sigma = \mathrm{cl}(\mathrm{ran}\ S(t)) \qquad X_\sigma^T = \mathrm{cl}(\mathrm{ran}\ S^T(t))$$

$$\mathcal{X}_\sigma = \mathrm{cl}(\mathrm{ran}\ \mathcal{S}(t)) \qquad \mathcal{X}_\sigma^T = \mathrm{cl}(\mathrm{ran}\ \mathcal{S}^T(t)).$$

Proof. Clearly, every generalized eigenspace is contained in the range of its corresponding semigroup operator for every $t \geqslant 0$.

Conversely, let $g \in M^q$ such that $g \perp X_\sigma$. Then it follows from Corollary 3.1.7 (iv) and Corollary 3.1.6 that $G^*g \in \mathcal{X}_0^T = \ker \mathcal{S}^T(t)$. We conclude that $G^*S^*(t)g = \mathcal{S}^T(t)G^*g = 0$ (Theorem 2.2.2) and hence $g \in \ker S^*(t) = (\mathrm{ran}\ S(t))^\perp$ (Lemma 2.2.1).

Secondly, let $g \in \mathcal{M}^q$ such that $\pi^T g \perp \mathcal{X}_\sigma$. Then it follows from Corollary 3.1.7 (ii) and Corollary 3.1.6 that $\mathcal{G}^*\pi^T g \in X_0^T = \ker S^T(t)$. We conclude that $\mathcal{G}^*\mathcal{S}^*(t)\pi^T g = S^T(t)\mathcal{G}^*\pi^T g = 0$ (Theorem 2.2.3) and hence $\pi^T g \in \ker \mathcal{S}^*(t) = (\mathrm{ran}\ \mathcal{S}(t))^\perp$ (Lemma 2.2.1). □

In order to prove the main result of this section, we need one more preliminary result concerning the relation between the small soltuions of Σ^T and those of Ω^T.

3.1.9 LEMMA. *Let* $z(t)$, $x(t)$ *be a small solution of* Σ^T *and define*

$$x(t) := -\int_t^{T_0} x(s)\,ds,\ t \geqslant -h.$$

Then $x(t) = 0$ *for* $t \geqslant T_0 - h$, *and* $\dot{x}(t) = L^T x_t + M^T \dot{x}_t$, $t \geqslant 0$.

Proof. It follows from Corollary 3.1.6 that $x(t) = 0$ for $t \geqslant T_0 - h$. This implies $z(t) = 0$ for $t \geqslant T_0$ and hence

$$z(t) = -\int_t^{T_0} L^T x_s ds = -\int_{-h}^0 d\eta^T(\tau) \int_t^{T_0} x(s+\tau)ds$$

$$= -\int_{-h}^0 d\eta^T(\tau) \int_{t+\tau}^{T_0} x(s)ds = L^T x_t,\ t \geqslant 0.$$

We conclude that $\dot{x}(t) = x(t) = z(t) + M^T x_t = L^T x_t + M^T \dot{x}_t,\ t \geqslant 0.$ □

The main result of this section follows. It is a generalization of a related result on retarded systems in the state space M^2 which has been proved by Manitius [93, Theorem 3.5.1].

3.1.10 THEOREM. *The following statements are equivalent.*

(i) *System* Σ *is complete.*
(ii) *System* Ω^T *has no nonzero small solution.*
(iii) $\ker F^* = \{0\}$.
(iv) *System* Ω *is complete.*
(v) *System* Σ^T *has no nonzero small solution.*
(vi) $\ker \mathcal{F}^* = \{0\}$.
(vii) *There is no nonzero* $\psi \in M^q$ *such that* $\langle\langle \psi, e^{\lambda \cdot} \rangle\rangle = 0$ *for every* $\lambda \in \mathbb{C}$.

3.1.11 REMARKS.

(i) Note that $\ker F^* = \{0\}$ if and only if the solutions $x(t)$ of Ω^T have the property

$$x(t) = 0\ \forall\ t \geqslant 0 \Rightarrow x(t) = 0\ \forall\ t \geqslant -h.$$

(ii) Note that $\ker \mathcal{F}^* = \{0\}$ if and only if the solutions $z(t)$, $x(t)$ of Σ^T have the property

$$x(t) = 0\ \forall\ t \geqslant 0 \Rightarrow x(t) = 0\ \forall\ t \geqslant -h.$$

PROOF OF THEOREM 3.1.10

'(i) $\Leftrightarrow$ (ii)' System Σ is complete if and only if $X_\sigma = M^p$ which means that $g \perp X_\sigma$ implies $g = 0$ for every $g \in M^q$. By Corollary 3.1.7 (iv), this is equivalent to $G^*g \in X_0^T \Rightarrow g = 0$. Now the equivalence of (i) and (ii) follows from the fact that $G^* : M^q \to W^{1,q}$ is bijective (Lemma 2.2.1).

'(ii) $\Leftrightarrow$ (iii)' System Ω^T has no nonzero small solution if and only if $\ker S^T(T_0) = X_0^T = \{0\}$ (Corollary 3.1.6). It follows from general semigroup theory that this is equivalent to $\ker S^T(h) = \{0\}$ and hence to $\ker F^* = \ker G^*F^* = \{0\}$ (Theorem 2.2.2).

'(ii) $\Leftrightarrow$ (v)' Lemma 3.1.9.

'(iv) $\Leftrightarrow$ (v)' System Ω is complete if and only if $X_\sigma = W^{1,p}$, which means that $\pi^T g \perp X_\sigma$ implies that $\pi^T g = 0$ for every $g \in M^q$. By Corollary 3.1.7 (ii), this is equivalent to $G^*\pi^T g \in X_0^T \Rightarrow \pi^T g = 0$. Now the equivalence of (iv) and (v) follows from the fact that $G^* : W^{-1,q} \to M^q$ is bijective (Lemma 2.2.1).

'(v) $\Leftrightarrow$ (vi)' System Σ^T has no nonzero small solution if and only if $\ker S^T(T_0) = X_0^T = \{0\}$ (Corollary 3.1.6). This is equivalent to $\ker F^* = \ker G^*F^* = \ker S^T(h) = \{0\}$ (Theorem 2.2.3).

'(vi) $\Leftrightarrow$ (vii)' Let $\psi \in M^q$ be given and note that $F^*\psi = \pi^T g \in W^{-1,q}$ for some $g \in M^q$ (Lemma 2.2.8 (iii)). Hence the equivalence of (vi) and (vii) follows from the fact that $F^*\psi = \pi^T g = 0$ if and only if

$$0 = \langle \pi^T g, e^{\lambda \cdot} \rangle = \langle F^*\psi, e^{\lambda \cdot} \rangle = \langle\langle \psi, e^{\lambda \cdot} \rangle\rangle$$

for every $\lambda \in \mathbb{C}$ (Lemma 2.1.5). □

COMPLETENESS IN THE STATE SPACE C

Every generalized eigenspace $X_\lambda \subset W^{1,p}$ can be regarded as a subspace of C. In this sense, X_λ is a generalized eigenspace of the semigroup $S_C(t)$ which is associated with the NFDE (2.1) in the state space C (see Remark 2.1.1). Let us define

$$C_\sigma = \mathrm{cl}_C(\mathrm{span}\{X_\lambda \mid \lambda \in \sigma(A)\}) \subset C.$$

We say that system Σ is *complete in* C if $C_\sigma = C$.

3.1.12 COROLLARY. *System Σ is complete in M^p if and only if it is complete in C.*

Proof. Let us regard $W^{1,p}$ and X_σ as subspaces of C. Then we have in any case $X_\sigma \subset C_\sigma$. If Σ is complete in M^p, then Ω is complete (Theorem 3.1.10) and hence $W^{1,p} = X_\sigma \subset C_\sigma$. Since $W^{1,p}$ is dense in C, this implies $C_\sigma = C$.

Conversely, let $C_\sigma = C$. Then $\{(\phi(0) - M\phi,\phi) | \phi \in C\} \subset X_\sigma$ and hence $X_\sigma = M^p$. □

MATRIX-TYPE CONDITIONS

Our next result is a computable completeness criterion for a rather general class of neutral systems. In the case of a single point delay, i.e. L and M are given by

$$L\phi = A_0\phi(0) + A_1\phi(-h), \ \phi \in C, \tag{3.1}$$

$$M\phi = A_{-1}\phi(-h), \ \phi \in C, \tag{3.2}$$

($A_0, A_1, A_{-1} \in \mathbb{R}^{n \times n}$), related results have been proved by Jakubczyk [62, Theorem 2], Bartosiewicz [9, Corollary 1], and O'Connor-Tarn [110, Theorem 4.1].

3.1.13 THEOREM. *Suppose that the equations*

$$\eta(\tau) = A_1 + \eta(-h), \ -h < \tau \leqslant \varepsilon\text{-}h, \tag{4.1}$$

$$\mu(\tau) = A_{-1} + \mu(-h), \ -h < \tau \leqslant \varepsilon\text{-}h, \tag{4.2}$$

hold for some $\varepsilon > 0$. *Then system* Σ *is complete if and only if the following equation holds for some* $\lambda \in \mathbb{C}$

$$\text{rank } [A_1 + \lambda A_{-1}] = n. \tag{5}$$

Proof. By Theorem 3.1.10, system Σ is complete if and only if Ω^T has no non-zero small solutions, which means that the implication

$$x(t) = 0 \ \forall \ t \geqslant \varepsilon\text{-}h \Rightarrow x(t) = 0 \ \forall \ t \geqslant -h \tag{6}$$

holds for every solution $x(t)$, $t \geqslant -h$, of Ω^T. Now let (4) be satisfied and define $x(t) := x(t-h)$, $f(t) := \dot{x}(t-h)$ for $0 \leqslant t \leqslant \varepsilon$. Then (6) is equivalent to

$$\left.\begin{array}{l} \dot{x}(t) = f(t), \ x(\varepsilon) = 0 \\ 0 = A_1^T x(t) + A_{-1}^T f(t) \end{array}\right\} \Rightarrow x(t) \equiv 0.$$

This means that

$$\operatorname{rank}\begin{bmatrix}\lambda I & -I \\ A_1 & A_{-1}\end{bmatrix} = n + \operatorname{rank}\begin{bmatrix}-I \\ A_{-1}\end{bmatrix} = 2n$$

for some $\lambda \in \mathbb{C}$ (see Appendix, Theorem A6). This is equivalent to (5). □

In the retarded case ($\mu(\tau) \equiv 0$), condition (6) reduces to

$$\operatorname{rank} A_1 = n. \tag{7}$$

This is precisely the completeness criterion which was derived by Banks-Manitius [8] (state space C) and Manitius [93] and Delfour-Manitius [29] (state space M^2).

Finally, let us briefly discuss the question under which conditions the operators F and $\mathcal{F}$ are bijective.

3.1.14 REMARK. Recall that $S(h) = GF$ and G is bijective. Hence F is bijective if and only if $S(t) : M^p \to M^p$ is a group. Correspondingly, $\mathcal{F}$ is bijective iff $\mathcal{S}(t) : W^{1,p} \to W^{1,p}$ is a group. Now recall that $S(t)$ and $\mathcal{S}(t)$ are isomorphic (Lemma 1.3.2 (iii)). Hence $S(t)$ is a group iff $\mathcal{S}(t)$ is a group. We conclude that F is bijective if and only if $\mathcal{F}$ is bijective.

Burns, Herdman and Stech [19, Theorem 2.4] have derived a matrix-type condition for $S(t)$ to be a group. Just for completeness, we present an alternative proof of their criterion. A related result on RFDEs can be found in Delfour-Manitius [29, Theorem 2.9].

3.1.15 PROPOSITION.

(i) *Suppose that*

$$\operatorname{rank} A_{-1} = n,\ A_{-1} = \lim_{\tau \downarrow -h} \mu(\tau) - \mu(-h) \in \mathbb{R}^{n \times n}. \tag{8}$$

Then the operator $F : M^p \to W^{-1,p}$ *is boundedly invertible.*

(ii) *If* $\mu(\tau)$ *is absolutely continuous with* L^q*-derivative on some interval* $(-h, \varepsilon-h]$, $\varepsilon > 0$, *then condition* (8) *is necessary and sufficient for bounded invertibility of* F.

Proof. We prove the corresponding result for the operator $F : W^{1,p} \to M^p$. For this sake let us define

$$\alpha(t) = \mu(t-h) - \int_0^t [\eta(s-h) - \eta(-h)]ds, \ 0 \leqslant t \leqslant h.$$

Moreover let $\phi \in W^{1,p}$ and $f \in M^p$. Then $F\phi = f$ if and only if $\phi(0) = f^0$ and

$$f^1(t-h) = \int_{-h}^{t-h} d\eta(\tau)\phi(\tau+h-t) + \int_{-h}^{t-h} d\mu(\tau)\dot{\phi}(\tau+h-t)$$

$$= [\eta(t-h)-\eta(-h)]\phi(0) - \int_{-h}^{t-h} [\eta(\tau)-\eta(-h)]\dot{\phi}(\tau+h-t)d\tau$$

$$+ \int_{-h}^{t-h} d\mu(\tau)\dot{\phi}(\tau+h-t)$$

$$= [\eta(t-h)-\eta(-h)]f^0 + \int_0^t d\alpha(s)\dot{\phi}(s-t)$$

for $0 \leqslant t \leqslant h$. Hence statement (i) follows from Theorem 1.1.4.

Now let $x \in \mathbb{R}^n$, $x \neq 0$, such that $x^T A_{-1} = 0$, and let $\mu(\tau)$ be absolutely continuous with L^q-derivative on some interval $(-h,\varepsilon-h]$, $\varepsilon > 0$. Then the above equation transforms into

$$f^1(t-h) = [\eta(t-h)-\eta(-h)]f^0 + A_{-1}\dot{\phi}(-t) + \int_0^t \dot{\alpha}(s)\dot{\phi}(s-t)ds$$

for $0 \leqslant t \leqslant \varepsilon$. Hence $x^T f^1(\tau)$ is continuous on the interval $[-h,\varepsilon-h]$ for every $f \in$ ran F with $f^0 = 0$ (Remark 1.1.1 (i)). We conclude that F is not surjective. □

OPEN PROBLEMS

The problem of finding a necessary and sufficient condition for the bounded invertibility of F is equivalent to that of finding a necessary and sufficient condition on $\alpha \in NBV([0,T);\mathbb{R}^{n\times n})$ such that the conclusions of Theorem 1.1.4 remain valid. This is not yet solved.

Also the problem of characterizing the injectivity of F, when (4) is not satisfied, is still open. If (4) is satisfied, then condition (5) shows that system Σ is complete if and only if Σ^T is complete. Hence by Theorem 3.1.10, system Σ is complete iff one and the same system (not the transposed!) has no nonzero small solution. In other words, $X_\sigma = M^p \Longleftrightarrow X_0 = \{0\}$. In general, this is an open problem.

One might also pose the question under which conditions the state space can be decomposed into a direct sum of the generalized eigenfunctions (X_σ) and the initial states of small solutions (X_0). This would mean $M^p = X_\sigma \oplus X_0$ □? Such a decomposition can obviously be obtained in the (extreme) case of a system with a finite spectrum. In general this problem is apparently not solved in the open literature on delay systems, even for retarded systems with a single point delay.

3.2 F-COMPLETENESS OF EIGENFUNCTIONS AND NONEXISTENCE OF NONTRIVIAL SMALL SOLUTIONS

It has been indicated by Manitius [93] (RFDE) that completeness in the sense of Section 3.1 might be a too restrictive property for delay systems. In particular, if the maximal delay does not appear on the right-hand side of each equation (componentwise), then the system cannot be complete. For example, consider the two-dimensional NFDE

$$\begin{aligned} \dot{x}_1(t) &= x_1(t-h) + \dot{x}_2(t-h) \\ \dot{x}_2(t) &= 0 \end{aligned} \tag{9}$$

which may be written in the form

$$d/dt\ (x(t) - A_{-1}x(t-h)) = A_0x(t) + A_1x(t-h)$$

where

$$A_0 = \begin{bmatrix} 0 & 0 \\ 0 & 0 \end{bmatrix},\ A_1 = \begin{bmatrix} 1 & 0 \\ 0 & 0 \end{bmatrix},\ A_{-1} = \begin{bmatrix} 0 & 1 \\ 0 & 0 \end{bmatrix}. \tag{10}$$

Obviously, condition (5) is not satisfied in this case. Also in the transposed situation - when the maximal delay does not occur in every state variable - completeness is impossible. Therefore it might be useful to work with a weaker notion of completeness. For retarded systems Manitius [91, 93] has introduced the concept of F-completeness. This has something to do with the completeness of eigenfunctions with respect to the dual state concept. We will extend these ideas to NFDEs in the state spaces M^p and $W^{1,p}$.

Let us first consider the system $\tilde{\Sigma}$ which is described by the semigroup $S^{T*}(t)$ on $W^{-1,p}$. By Lemma 2.4.3, the generalized eigenspaces of the generator A^{T*} are given by $FX_\lambda \subset W^{-1,p}$. These eigenspaces cannot span the whole state space $W^{-1,p}$ unless ran F is dense in this space. A suitable candidate for the closed span might be cl(ran F). An analogous situation is given in the case of system $\tilde{\Omega}$. The generalized eigenspaces of the corresponding semigroup $S^{T*}(t)$ are given by $\mathcal{F}X_\lambda \subset M^p$ (Lemma 2.4.3). The closed span of these eigenspaces will be studied in the closure of ran $\mathcal{F}$.

3.2.1 DEFINITION.

System Σ *is said to be* F-*complete if* $\text{cl}(FX_\sigma) = \text{cl}(\text{ran } F)$.
System Ω *is said to be* $\mathcal{F}$-*complete if* $\text{cl}(\mathcal{F}X_\sigma) = \text{cl}(\text{ran } \mathcal{F})$.

This concept of F-completeness is obviously weaker than completeness in the sense of Definition 3.1.1. It is related to the 'triviality' of small solutions which is defined as follows.

3.2.2 DEFINITION. *A small solution* w(t), x(t) *of system* Σ *is said to be trivial if* $x(t) = 0$ *for every* $t \geqslant 0$.

A small solution $x(t)$ *of system* Ω *is said to be trivial if* $x(t) = 0$ *for every* $t \geqslant 0$.

In other words, a small solution of Σ or Ω is trivial in the above sense iff the initial state in the dual state concept is zero.

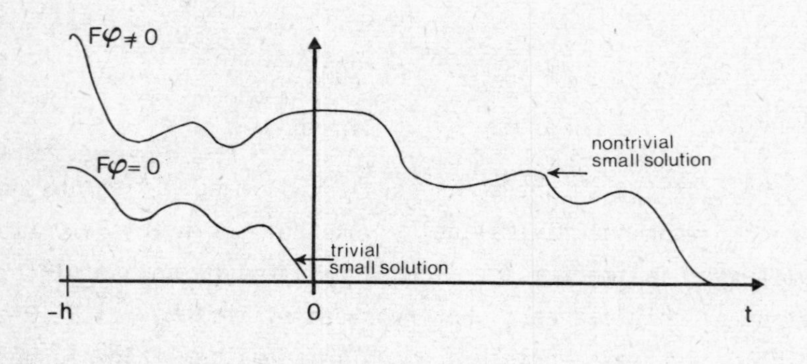

Figure 3

The following theorem is the main result of this section. For retarded systems in the state space M^2 it has been proved by Manitius [93, Theorem 5.6].

3.2.3 THEOREM. *The following statements are equivalent.*

(i) *System* Σ *is* F*-complete.*

(ii) *System* Ω^T *has only trivial small solutions, i.e.* $X_0^T = \ker F^*$.

(iii) *For every* $\psi \in W^{1,q}$ *the following implication holds*

$$F^*G^*F^*\psi = 0 \Rightarrow F^*\psi = 0. \tag{11}$$

(iv) *System* Ω *is* F*-complete.*

(v) *System* Σ^T *has only trivial small solutions, i.e.* $X_0^T = \ker F^*$.

(vi) *For every* $\psi \in M^q$ *the following implication holds*

$$F^*G^*F^*\psi = 0 \Rightarrow F^*\psi = 0. \tag{12}$$

(vii) *If the complex function* $\langle\langle\psi,e^{s\cdot}\rangle\rangle\,\Delta(s)^{-1}$, $s \in \mathbb{C}$, *is entire for some* $\psi \in M^q$, *then* $\langle\langle\psi,e^{s\cdot}\rangle\rangle = 0$ *for every* $s \in \mathbb{C}$.

3.2.4 REMARKS.

(i) Condition (11) is equivalent to the property

$$x(t) = 0 \;\forall\, t \geq h \Rightarrow x(t) = 0 \;\forall\, t \geq 0$$

for the solutions of Ω^T.

(ii) Condition (12) is equivalent to the property

$$x(t) = 0 \;\forall\, t \geq h \Rightarrow x(t) = 0 \;\forall\, t \geq 0$$

for the solutions of Σ^T.

PROOF OF THEOREM 3.2.3

'(i) $\Leftrightarrow$ (ii)' System Σ is F-complete if and only if $F^*\psi \perp X_\sigma \Rightarrow F^*\psi = 0$ for every $\psi \in W^{1,q}$. Moreover it follows from Corollary 3.1.7 that $F^*\psi \perp X_\sigma$ if and only if $\psi \in X_0^T$.

'(ii) $\Leftrightarrow$ (iii)' By Corollary 3.1.6 and Theorem 2.2.2, we have

$$X_0^T = \ker S^T(nh) = \ker (F^*G^*)^{n-1}F^*.$$

Hence the equivalence of (ii) and (iii) follows easily by induction (see also Remark 3.2.4 (i)). The equivalence of (ii) and (v) follows from Lemma 3.1.9.

The equivalence of (iv), (v), and (vi) can be proved by analogous arguments.

'(v) $\Longleftrightarrow$ (vii)' In the proof of Theorem 3.1.10, we have seen that $F^*\psi = 0$ if and only if $\langle\langle \psi, e^{s\cdot}\rangle\rangle = 0$ for every $s \in \mathbb{C}$ ($\psi \in M^q$). On the other hand it follows from Theorem 2.4.10 (iii) that $\psi \in X_0^T$ if and only if the complex function $\langle F^*\psi, e^{s\cdot}\rangle\, \Delta(s)^{-1} = \langle\langle \psi, e^{s\cdot}\rangle\rangle\, \Delta(s)^{-1}$, $s \in \mathbb{C}$, is entire (see also Theorem 3.1.4). This proves the equivalence of (v) and (vii). □

The following matrix-type condition for systems with a single point delay is a special case of Theorem 4.3.7 in the next chapter. It can be extended to systems with commensurable delays. However, in a more general situation, the derivation of an analogous result seems to be a hard problem.

3.2.5 <u>COROLLARY</u>. *Let* L *and* M *be given by* (3). *Then system* Σ *is* F-*complete if and only if*

$$\max_{\lambda\in\mathbb{C}} \operatorname{rank} \begin{bmatrix} A_0-\lambda I & A_1+\lambda A_{-1} \\ A_1+\lambda A_{-1} & 0 \end{bmatrix} = n + \max_{\lambda\in\mathbb{C}} \operatorname{rank}\, [A_1+\lambda A_{-1}] \tag{13}$$

In the retarded case ($A_{-1} = 0$) condition (13) reduces to

$$\max_{\lambda\in\mathbb{C}} \operatorname{rank} \begin{bmatrix} A_0-\lambda I & A_1 \\ A_1 & 0 \end{bmatrix} = n + \operatorname{rank} A_1. \tag{14}$$

This is precisely the criterion for F-completeness in the state space M^2 which has been derived by Manitius [93, Corollary 6.4].

3.2.6 <u>EXAMPLES</u>

(i) Consider the two-dimensional system

$$\begin{aligned} \dot{x}_1(t) &= x_1(t-h) + \dot{x}_2(t-h) \\ \dot{x}_2(t) &= \alpha x_1(t) \end{aligned} \tag{15}$$

which is described by the matrices

$$A_0 = \begin{bmatrix} 0 & 0 \\ \alpha & 0 \end{bmatrix}, \; A_1 = \begin{bmatrix} 1 & 0 \\ 0 & 0 \end{bmatrix}, \; A_{-1} = \begin{bmatrix} 0 & 1 \\ 0 & 0 \end{bmatrix}. \tag{16}$$

In this case we have

$$\text{rank} \begin{bmatrix} A_0-\lambda I & \lambda A_1 + A_{-1} \\ A_1+\lambda A_{-1} & 0 \end{bmatrix} = \text{rank} \begin{bmatrix} -\lambda & 0 & 1 & \lambda \\ \alpha & -\lambda & 0 & 0 \\ 1 & \lambda & 0 & 0 \\ 0 & 0 & 0 & 0 \end{bmatrix},$$

and hence condition (13) is satisfied if and only if $\alpha \neq -1$. In particular, the introductory example (9) ($\alpha = 0$) is F-complete.

(ii) The scalar n-th order differential-difference equation

$$z^{(n)}(t) = \sum_{j=0}^{n-1} \alpha_j z^{(j)}(t) + \sum_{j=0}^{n} \beta_j z^{(j)}(t-h) \tag{17}$$

can be rewritten as a first order system of neutral type ($x_j(t) := z^{(j-1)}(t)$ for $j = 1,\ldots,n$) where

$$A_0 = \begin{bmatrix} 0 & 1 & & \\ & \ddots & \ddots & \\ & & 0 & 1 \\ \alpha_0 & \cdots & & \alpha_{n-1} \end{bmatrix}, \tag{18.1}$$

$$A_1 = \begin{bmatrix} 0 & \cdots & 0 \\ \vdots & & \vdots \\ 0 & \cdots & 0 \\ \beta_0 & \cdots & \beta_{n-1} \end{bmatrix}, \; A_{-1} = \begin{bmatrix} 0 & \cdots & 0 \\ \vdots & & \vdots \\ 0 & \cdots & 0 \\ 0 & \cdots\; 0 & \beta_n \end{bmatrix}. \tag{18.2}$$

This system is not complete unless $n = 1$. However, some elementary operations show that the rank of the matrix

$$\begin{bmatrix} A_0-\lambda I & A_1+\lambda A_{-1} \\ A_1+\lambda A_{-1} & 0 \end{bmatrix}$$

$$
= \begin{bmatrix}
-\lambda & & 1 & & 0 & \cdots & 0 \\
& \ddots & & \ddots & \vdots & & \vdots \\
& & -\lambda & 1 & 0 & \cdots & 0 \\
\alpha_0 & \cdots & \alpha_{n-2} & \alpha_{n-1}-\lambda & \beta_0 \cdots \beta_{n-2} & & \beta_{n-1}+\lambda\beta_n \\
0 & \cdots & \cdots & 0 & 0 & \cdots & 0 \\
\vdots & & & \vdots & \vdots & & \vdots \\
0 & \cdots & \cdots & 0 & \vdots & & \vdots \\
\beta_0 & \cdots & \beta_{n-2} & \beta_{n-1}+\lambda\beta_n & 0 & \cdots & 0
\end{bmatrix}
$$

coincides with the rank of

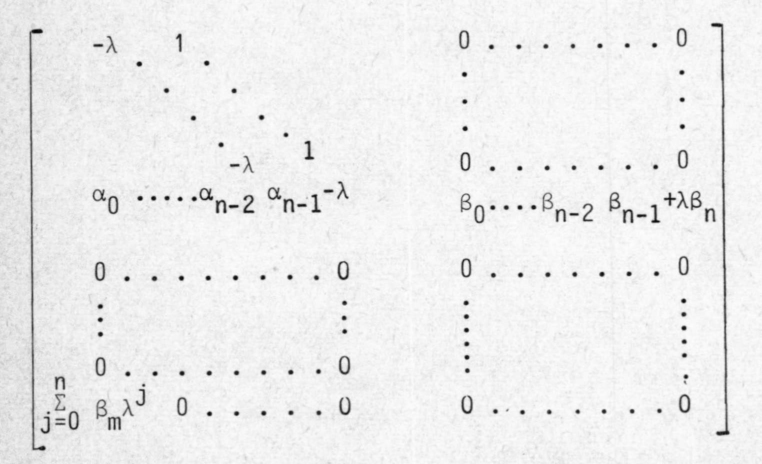

The maximal rank of this matrix (over $\lambda \in \mathbb{C}$) is n, if all β_j are zero, and n+1 otherwise. Hence (13) is satisfied in both cases. We conclude that system (17), respectively (18), is always F-complete.

4 Controllability and observability

In this chapter we deal with controllability and observability properties of neutral systems in the state spaces M^p and $W^{1,p}$. This will be done within the functional analytic context developed in Chapter 2. Our work in this area has been mainly influenced by two recent papers of Manitius [94, 95] on approximate controllability of linear retarded systems in the state space M^2. Other contributions to the approximate function space controllability of RFDEs can be found in Olbrot [114, 117], Kurcyusz-Olbrot [79], Manitius-Triggiani [98-100], Manitius [90-92], Pandolfi [124], Korytowski [75], Popov [129], Zmood [151], Choudhury [21], Delfour-Mitter [31], Minjuk [105], Minjuk-Stepanjuk [106] and Salamon [134].

Observability properties of retarded systems have been investigated, for instance, by Olbrot [115, 116, 119], Lee [82], Lee-Olbrot [83], Kwong [81], Kociecki [73] and Kopeikina-Mularthik [74].

Most of the earlier work in this area has been done on spectral controllability and observability (Krasovskii-Kurzhanskii [77], Krasovskii-Osipov [78], Osipov [123], Pandolfi [125], Bhat-Koivo [13]) and on Euclidean controllability (Kirillova-Curakova [69], Gabasov-Kirillova [38], Zmood [151, 152], Manitius-Olbrot [96]). For systems with control delays only, we refer to Chyung [22], Sebakhy-Bayoumi [139], Olbrot [112], Banks-Jacobs-Latina [6], Manitius-Olbrot [97], Kwon-Pearson [80], Klamka [70, 71] and Lewis [87].

We will not go into the algebraic concepts of controllability and observability which have been developed within the theory of systems over rings. The interested reader is referred to Kamen [65, 66], Sontag [141], Morse [108], Zakian-Williams [150], Olbrot-Zak [121, 122], Lee-Olbrot [83, 84], Jakubczyk-Olbrot [64], Hautus-Sontag [44] and Hazewinkel [45].

Controllability properties of NFDEs have been mainly analyzed in the state space $W^{1,p}$ and for systems of the form

$$d/dt(x(t) - A_{-1}x(t-h)) = A_0x(t) + A_1x(t-h) + B_0u(t).$$

There has been a series of papers on exact controllability for this class of systems, namely Banks-Jacobs-Langenhop [3-5], Jacobs-Langenhop [60, 61], Rodas-Langenhop [130], Jakubczyk [62], and also Bartosiewicz [10, Proposition 16], O'Connor-Tarn [110, Corollary 5.8]. Recently, Bartosiewicz [10] and O'Connor [109] have studied - independently and by different methods - the approximate controllability of general neutral systems in the state space $W^{1,p}$. Again in the case of NFDEs with a single point delay, they have derived a computable rank criterion in terms of the system matrices. Bartosiewicz [10] also allows delays in the control variable.

It is interesting to re-examine these properties of neutral systems by means of the approach of Chapter 2. In particular, we obtain duality results for systems with general delays in input, state and output. These have been stated as open problems in O'Connor [109] and Ito [59]. Moreover, we extend the controllability criterion of Bartosiewicz [10] and O'Connor [109] to a rather general class of neutral systems (Section 4.2). Also, we introduce the weaker concept of F-controllability for neutral systems with general delays in state and input (Section 4.3). A duality result for this controllability concept is obtained and - in the case of a single point delay - a rank condition. A preliminary section is devoted to the well-known basic concepts of spectral controllability and observability.

Throughout this chapter, we denote by Σ, Ω, Σ^T, Ω^T the control systems of Section 2.3 and by Σ_λ, Σ_λ^* $(\lambda \in \sigma(A))$ the projected systems which are described in Section 2.4.

4.1 SPECTRAL CONTROLLABILITY AND OBSERVABILITY

The concepts of spectral controllability and observability of a retarded system were first introduced - without the explicit use of these expressions - for the purpose of stabilization in a series of Russian papers in the mid-1960s (see, e.g., Krasovskii-Osipov [78], Krasovskii [76], Osipov [123], Krasovskii-Kurzhanskii [77]). Later, Pandolfi [125, 126] and Bhat-Koivo [13] derived independently a criterion for spectral controllability and observability of retarded systems with undelayed input and output. This has been extended to retarded systems with output delays (Salamon [132]) and to neutral systems with input delays (Bartosiewicz [10, Theorem 4]). In [10], the input is not included in the spectral projection operator. This leads to a finite-dimensional projected system with input delays. On the basis of results of Section 2.4, we give a slightly different (but equivalent) definition of spectral controllability.

4.1.1 <u>DEFINITION.</u>

(i) *System* Σ *(and* Ω *in the case* $\Gamma = 0$*) is said to be spectrally controllable if* Σ_λ *is controllable for all* $\lambda \in \sigma(A)$.

(ii) *System* Ω^T *(and* Σ^T *in the case* $\Gamma = 0$*) is said to be spectrally observable if* Σ_λ^* *is observable for all* $\lambda \in \sigma(A)$.

The following rank criterion shows that our definition of spectral controllability coincides with that of Bartosiewicz [10] ($\Gamma = 0$).

4.1.2 <u>PROPOSITION.</u> *Let* $\lambda \in \sigma(A)$ *be given. Then* Σ_λ *is controllable* (Σ_λ^* *is observable) if and only if*

$$\text{rank}\,[\Delta(\lambda),\ B(e^{\lambda\cdot}) + \lambda\Gamma(e^{\lambda\cdot})] = n. \tag{1}$$

<u>Proof</u>. Let $A_\lambda \in \mathbb{C}^{N\times N}$ and $B_\lambda \in \mathbb{C}^{N\times m}$ be defined as in Section 2.4. Then it follows from the Hautus condition that Σ_λ is not controllable if and only if rank $[\lambda I - A_\lambda, B_\lambda] < N$. This means that

$$x^*[\lambda I - A_\lambda] = 0,\ x^*B_\lambda = 0,\ x \neq 0,$$

for some $x \in \mathbb{C}^N$. Equivalently, $\psi = \Psi x \in W^{1,q}$ satisfies

$$(\bar{\lambda}I - A^T)\psi = 0,\ \int_{-h}^{0} \psi^*(\tau)d\beta(\tau) + \int_{-h}^{0} \dot{\psi}^*(\tau)d\gamma(\tau) = 0,\ \psi \neq 0, \tag{2}$$

(see equation (2.66)). Now it follows from Lemma 2.4.1 that $\psi \in \ker(\bar{\lambda}I - A^T)$ if and only if $\psi(\tau) = e^{\bar{\lambda}\tau}\psi(0)$ for $-h \leqslant \tau \leqslant 0$ and $\Delta^*(\lambda)\psi(0) = 0$. Hence (2) is equivalent to

$$\psi^*(0)\Delta(\lambda) = 0,\ \psi^*(0)[B(e^{\lambda\cdot}) + \lambda\Gamma(e^{\lambda\cdot})] = 0,\ \ \psi(0) \neq 0.$$

This means that (1) is not satisfied. □

At first glance, (1) seems to be a rather unhandy criterion for spectral controllability since it has to be satisfied for every $\lambda \in \sigma(A)$. This is in general an infinite set and impossible to compute completely. However, in many cases condition (1) can be checked directly without computation of any eigenvalue, just by looking long enough at the matrix. Moreover, for retarded systems of the form

$$\dot{x}(t) = A_0x(t) + A_1x(t-h) + B_0u(t) \tag{3}$$

some research effort has been directed toward transforming condition (1) into one which is easier to handle (Manitius-Triggiani [98]). These ideas have been generalized to neutral systems in O'Connor [109] and O'Connor-Tarn [110].

In the rest of this section we will focus on the important question of how spectral controllability (observability) is related to the function space controllability (observability) properties of neutral systems.

THE REACHABLE AND THE UNOBSERVABLE SUBSPACE

The reachable subspaces associated with the systems Σ and Ω are given by

$$R_t = \{(w(t),x_t,u_t) \in M^p \times L^p | w(\cdot),\ x(\cdot) \text{ is a solution of } \Sigma,\ (2.42) \text{ corresponding to some input } u(\cdot) \in L^p([0,t];\mathbb{R}^m) \text{ and the initial state } \phi = 0,\ \xi = 0\},$$

$$\mathcal{R}_t = \{(x_t,u_t) \in W^{1,p} \times L^p | x(\cdot) \text{ is a solution of } \Omega,\ (2.47) \text{ corresponding to some input } u(\cdot) \in L^p([0,t];\mathbb{R}^m) \text{ and the initial state } \phi = 0,\ \xi = 0\}$$

for $t \geqslant 0$. By analogy, we introduce the unobservable subspaces of the systems Ω^T and Σ^T as follows

$$N_t^T = \{\psi \in M^q | \text{the output } y(\cdot) \text{ of } \Sigma^T,\ (2.8) \text{ vanishes on the interval } [0,t]\},$$

$$\mathcal{N}_t^T = \{\psi \in W^{1,q} | \text{the output } y(\cdot) \text{ of } \Omega^T,\ (2.11) \text{ vanishes on the interval } [0,t]\}.$$

Moreover, we define

$$R = \bigcup_{t>0} R_t \qquad N^T = \bigcap_{t>0} N_t^T$$

$$\mathcal{R} = \bigcup_{t>0} \mathcal{R}_t \qquad \mathcal{N}^T = \bigcap_{t>0} \mathcal{N}_t^T.$$

If necessary, these subspaces will be interpreted as their obvious complex extensions.

Let us now summarize some basic properties of the reachable and the unobservable subspaces which will be needed frequently.

4.1.3 REMARKS

(i) The subspace

$$\tilde{R}_t = [F\ E]R_t = \{F\phi + E\xi \mid (\phi,\xi) \in R_t\} \subset W^{-1,p}$$

is precisely the space of all final states $\pi(w(t),w^t,x^t) \in W^{-1,p}$ of $\tilde{\Sigma}$ which are reachable from zero via some control function $u(\cdot) \in L^p([0,t];\mathbb{R}^m)$.

(ii) The subspace

$$\tilde{\mathcal{R}}_t = [\mathcal{F}\ \mathcal{E}]\mathcal{R}_t = \{\mathcal{F}\phi + \mathcal{E}\xi \mid (\phi,\xi) \in \mathcal{R}_t\} \subset M^p$$

is precisely the space of all final states $(x(t),x^t) \in M^p$ of $\tilde{\Omega}$ which are reachable from zero via some control function $u(\cdot) \in L^p([0,t];\mathbb{R}^m)$.

(iii) Let $\Gamma = 0$ and let $\phi \in W^{1,p}$, $\xi \in L^p$ be given. Then it follows from Remark 2.3.1 that $(\phi,\xi) \in \mathcal{R}_t$ iff $(\iota\phi,\xi) \in R_t$.

(iv) Let $\Gamma = 0$ and let $\psi \in W^{1,q}$ be given. Then it follows from Remark 2.3.3 that $\psi \in \mathcal{N}_t^T$ iff $\iota^T\psi \in N_t^T$.

(v) Let $R_\lambda \subset \mathbb{C}^N$ be the reachable subspace of Σ_λ. Moreover let $\Psi \in W^{1,q}([-h,0];\mathbb{C}^{n\times N})$ be as in Section 2.4. Then it follows from (i) and Proposition 2.4.8 that $R_\lambda = \{\langle\Psi,F\phi + E\xi\rangle \mid (\phi,\xi) \in R\}$.

(vi) Let $N_\lambda^* \in \mathbb{C}^N$ be the unobservable subspace of system Σ_λ^*. Then it follows from the spectral projection result on system Ω^T that $N_\lambda^* = \{x \in \mathbb{C}^N \mid \Psi x \in \mathcal{N}^T\}$.

The last two statements in the above remark already describe a basic relation between the spectral and function space concepts of controllability and observability. For the derivation of some consequences of these facts, it is convenient to make use of the duality relations between the reachable and the unobservable subspaces. These preliminary results are crucial for the whole theory of Chapter 4.

DUALITY

The duality relations between the reachable and the unobservable subspaces are described by means of the structural operators.

4.1.4 LEMMA

(i) *Let* $\psi \in W^{1,q}$, $g \in M^q$, $d \in L^q$ *be given. Then*

$$(F^*\psi, E^*\psi) \perp R_t \iff \psi \in \mathcal{N}_t^T \quad (t \geq 0).$$

$$(g,d) \perp R_t \iff G^*g \in \mathcal{N}_{t-h}^T,\ d = -D^*G^*g \ (t \geq h).$$

(ii) *Let* $\psi \in M^q$, $g \in M^q$, $d \in L^q$ *be given. Then*

$$(F^*\psi, E^*\psi) \perp \mathcal{R}_t \iff \psi \in N_t^T \quad (t \geq 0),$$

$$(\pi^T g, d) \perp \mathcal{R}_t \iff G^*\pi^T g \in N_{t-h}^T,\ d = \mathcal{D}^*G^*\pi^T g \ (t \geq h).$$

Proof. It suffices to prove (i) since the proof of (ii) is strictly analogous.

First note that $(F^*\psi, E^*\psi) \perp R_t$ iff $\psi \perp [F\ E]\, R_t = \tilde{R}_t$ (Remark 4.1.3 (i)). Moreover let $y(\cdot)$ be the output of Ω^T which corresponds to the initial state $\psi \in W^{1,q}$. Then it follows from Theorem 2.3.5 that $\psi \perp \tilde{R}_t$ if and only if $\int_0^t y^T(t-s)u(s)ds = 0$ for every $u(\cdot) \in L^p([0,t];\mathbb{R}^m)$. This means that $\psi \in \mathcal{N}_t^T$.

Secondly note that, by Proposition 2.3.9 (i), we have

$$R_t = \{(G[F\phi + E\xi],0) \mid (\phi,\xi) \in R_{t-h}\} + \{(GD\zeta,\zeta) \mid \zeta \in L^p\}$$

for every $t \geq h$. Hence $(g,d) \perp R_t$ if and only if $g \perp G[F\ E]R_{t-h}$ and

$$\langle g, GD\zeta\rangle_{M^q,M^p} + \langle d,\zeta\rangle_{L^q,L^p} = 0$$

for every $\zeta \in L^p$. This is equivalent to $(F^*G^*g, E^*G^*g) \perp R_{t-h}$, $D^*G^*g + d = 0$, and hence to $G^*g \in \mathcal{N}_{t-h}^T$, $d = -D^*G^*g$. □

4.1.5 REMARK. The duality relations of Lemma 4.1.4 remain valid, if the finite-time subspaces R_t, $\mathcal{R}_t$, $\mathcal{N}_t^T$, N_t^T are replaced by the infinite-time subspaces R, $\mathcal{R}$, $\mathcal{N}^T$, N^T.

SPECTRAL AND FUNCTION SPACE PROPERTIES

We are now going to lay the ground for the relation between spectral controllability (observability) and function space controllability (observability).

4.1.6 LEMMA. *Let $\lambda \in \sigma(A)$ be given. Then the following statements are equivalent.*

(i) Σ_λ *is controllable.*

(ii) Σ_λ^* *is observable.*

(iii) $X_{\bar\lambda}^T \cap N^T = \{0\}$.

(iv) $N^T \subset X^{\bar\lambda^T}$.

(v) $(\phi,0) \in cl(R)$ *for every* $\phi \in X_\lambda$.

(vi) $FX_\lambda \subset cl([F\ E]\ R)$.

In the case $\Gamma = 0$ *the following statements are equivalent to* (i), (ii), (iii), (iv), (v), *and* (vi).

(vii) $X_{\bar\lambda}^T \cap N^T = \{0\}$.

(viii) $N^T \subset X^{\bar\lambda^T}$.

(ix) $(\phi,0) \in cl(R)$ *for every* $\phi \in X_\lambda$.

(x) $FX_\lambda \subset cl([F\ E]\ R)$.

Proof. Clearly, (i) is equivalent to (ii). Moreover it follows from Remark 4.1.3 (vi) that $N_\lambda^* = \{0\}$ if and only if $\Psi x \in N$ implies $x = 0$ for every $x \in \mathbb{C}^N$. This means that $X_{\bar\lambda}^T \cap N^T = \{0\}$. Hence (ii) and (iii) are equivalent.

'(iii) ⇒ (iv)' First note that N^T is invariant under the semigroup $S^T(t)$ and that the resolvent operator $(sI-A^T)^{-1}$ is given by

$$(sI-A^T)^{-1}\psi = \int_0^\infty e^{-st}S^T(t)\psi dt, \quad \psi \in W^{1,q},$$

if Re s is sufficiently large. By analyticity, this implies that N^T is invariant under $(sI-A^T)^{-1}$ for every $s \notin \sigma(A^T)$. Hence the formula in Remark 2.4.5 (iv) shows that N^T is also invariant under the projection operator $P_{\bar\lambda}^T$.

Now let $\psi \in N^T$. Then $P_{\bar\lambda}^T\psi \in X_{\bar\lambda}^T \cap N^T$ and hence, by (iii), $\psi \in \ker P_{\bar\lambda}^T = X^{\bar\lambda^T}$. This proves (iv).

'(iv) ⇒ (v)' Let $(g,d) \perp R$. Then, by Lemma 4.1.4, we have $G^*g \in N^T \subset X^{\bar\lambda^T}$. By Theorem 2.4.6, this implies that $g \perp X_\lambda$. Hence the pair (g,d) is orthogonal to $(\phi,0)$ for every $\phi \in X_\lambda$.

'(v) ⇒ (vi)' This implication is trivial.

'(vi) ⇒ (iii)' Let $\psi \in X_{\bar\lambda}^{T} \cap N^{T}$. Then, by Lemma 4.1.4, $(F^*\psi, E^*\psi) \perp R$ and hence $F^*\psi \perp X_{\lambda}$. By Theorem 2.4.6, this implies that $\psi \in X^{\bar\lambda^T}$ and thus $\psi = 0$.

Thus we have proved the equivalence of the statements (i), (ii), (iii), (iv), (v), and (vi).

'(iii) ⟺ (vii)' Let $\Gamma = 0$ and $\psi \in W^{1,q}$. Then it follows from Corollary 2.4.4 and Remark 4.1.3 (iv) that $\psi \in X_{\bar\lambda}^{T} \cap N^{T}$ if and only if $\iota^{T}\psi \in X_{\bar\lambda}^{T} \cap N^{T}$.

'(vii) ⇒ (viii) ⇒ (ix) ⇒ (x) ⇒ (vii)' This can be proved with precisely the same arguments as the implications '(iii) ⇒ (iv) ⇒ (v) ⇒ (vi) ⇒ (iii)' above. □

The equivalence of the statements (i) and (ix) in the previous lemma has been proved by Bartosiewicz [10, Theorem 5].

Now recall that the following equations hold for every $t \geq T_0$

$$\mathrm{cl}(\mathrm{ran}\ S(t)) = \mathrm{cl}(\mathrm{span}\ \{X_{\lambda}\}),\ \ker S^{T}(t) = \bigcap_{\lambda} X^{\lambda^T},$$

$$\mathrm{cl}(\mathrm{ran}\ \mathcal{S}(t)) = \mathrm{cl}(\mathrm{span}\ \{X_{\lambda}\}),\ \ker \mathcal{S}^{T}(t) = \bigcap_{\lambda} X^{\lambda^T},$$

(Corollary 3.1.6 and Proposition 3.1.8).

Combining these facts with Lemma 4.1.6, we obtain a very useful characterization of spectral controllability and observability, namely the following result.

4.1.7 <u>THEOREM</u>. *For every* $t \geq T_0$ *the following statements are equivalent.*

(i) Σ *is spectrally controllable.*

(ii) Ω^{T} *is spectrally observable.*

(iii) $(S(t)\phi, 0) \in \mathrm{cl}(R)$ *for every* $\phi \in M^{p}$.

(iv) $N^{T} \subset \ker S^{T}(t)$.

In the case $\Gamma = 0$ *the following statements are equivalent to* (i), (ii), (iii), *and* (iv).

(v) Ω *is spectrally controllable.*

(vi) Σ^{T} *is spectrally observable.*

(vii) $(\mathcal{S}(t)\phi, 0) \in \mathrm{cl}(\mathcal{R})$ *for every* $\phi \in W^{1,p}$.

(viii) $\mathrm{N}^{T} \subset \ker \mathrm{S}^{T}(t)$.

Note that Theorem 4.1.7 also makes sense for the real subspaces R, $\mathcal{R}$, $\mathcal{N}^T$, N^T, whereas for the formulation of Lemma 4.1.6 we need the complex extensions of these spaces.

APPROXIMATE NULL-CONTROLLABILITY AND FINAL OBSERVABILITY

At the end of this section we show that spectral controllability (observability) is equivalent to approximate null-controllability (final observability). These notions are defined as follows.

4.1.8 DEFINITION

(i) *System* Σ *is said to be approximately null-controllable in time* $t > h$ *if for every* $\phi \in M^p$, $\xi \in L^p$ *and every* $\varepsilon > 0$ *there exists an input function* $u \in L^p([0,t];\mathbb{R}^m)$ *such that the corresponding solution* $w(t)$, $x(t)$ *of* Σ, (2.42), *satisfies* $\|(w(t),x_t)\|_{M^p} < \varepsilon$ *and* $\|u_t\|_{L^p} < \varepsilon$.

(ii) *System* Ω *is said to be approximately null-controllable in time* $t > h$ *if for every* $\phi \in W^{1,p}$, $\xi \in L^p$, *and every* $\varepsilon > 0$ *there exists an input function* $u \in L^p([0,t];\mathbb{R}^m)$ *such that the corresponding solution* $x(t)$ *of* Ω, (II.47), *satisfies* $\|x_t\|_{W^{1,p}} < \varepsilon$ *and* $\|u_t\|_{L^p} < \varepsilon$.

(iii) *System* Ω^T *is said to be finally observable in time* $t > h$ *if the solutions* $x(\cdot)$ *of* Ω^T *satisfy*

$$y(s) = 0,\ 0 \leqslant s \leqslant t \Rightarrow x(s) = 0,\ s \geqslant t-h.$$

(iv) *System* Σ^T *is said to be finally observable in time* $t > h$ *if the solutions* $x(\cdot)$ *of* Σ^T *satisfy*

$$y(s) = 0,\ 0 \leqslant s \leqslant t \Rightarrow x(s) = 0,\ s \geqslant t-h.$$

4.1.9 REMARKS

(i) System Σ is approximately null-controllable in time $t > h$ iff $(S(t-h)G[F\phi + E\xi],0) \in \mathrm{cl}(R_t)$ for every $\phi \in M^p$ and every $\xi \in L^p$ (Proposition 2.3.9 (i)).

(ii) System Ω is approximately null-controllable in time $t > h$ iff $(\mathcal{S}(t-h)\mathcal{G}[\mathcal{F}\phi + \mathcal{E}\xi],0) \in \mathrm{cl}(\mathcal{R}_t)$ for every $\phi \in W^{1,p}$ and every $\xi \in L^p$ (Proposition 2.3.9 (ii)).

(iii) System Ω^T is finally observable in time $t > h$ if and only if $N_t^T \subset \ker S^T(t)$.

(iv) System Σ^T is finally observable in time $t > h$ if and only if $N_t^T \subset \ker S^T(t)$.

For the proof of the desired equivalence it remains to show that the unobservable subspaces N_t^T, N_t^T do not decrease and that the closure of the reachable subspaces R_t, R_t does not increase with time. The rest of the job is done by Theorem 4.1.7.

4.1.10 LEMMA. *There exists a (minimal) time* $T_1 \leqslant (n+1)h$ *such that*

$$cl(R) = cl(R_{t+h}), \quad N^T = N_t^T,$$

$$cl(R) = cl(R_{t+h}), \quad N^T = N_t^T$$

for every $t > T_1$.

Proof. Recall that the operator $G^* : M^q \to W^{1,q}$ is bijective (Lemma 2.2.1). Hence it follows from Lemma 4.1.4 that $N^T = N_t^T$ if and only if $cl(R) = cl(R_{t+h})$.

Now let $T_1 > 0$ such that the exponential growth of the entire functions

$$\det \Delta(s), \ \mathrm{adj}\ \Delta(s)\ [B(e^{s\cdot}) + s\Gamma(e^{s\cdot})], \ s \in \mathbb{C},$$

is less than or equal to $T_1 - h$. Note that we can choose $T_1 \leqslant (n+1)h$.

Moreover, let $\psi \in N_t^T$ for some $t > T_1$ and define $y(\tau) = 0$ for $0 \leqslant \tau \leqslant t$ and it follows from Proposition 2.4.9 that the Laplace transform $\hat{y}(s)$ of $y(\tau)$, $\tau \geqslant 0$, satisfies the following equation

$$\det \Delta(s)\ \hat{y}(s) = [B^T(e^{s\cdot}) + s\Gamma^T(e^{s\cdot})]\ \mathrm{adj}\ \Delta^T(s)\langle F^*\psi, \iota e^{s\cdot}\rangle^T$$

$$+ \det \Delta(s) \left(B^T(e^{s\cdot} * \psi) + s\Gamma^T(e^{s\cdot} * \dot{\psi} - e^{s\cdot}\psi(0))\right)$$

$$=: g(s).$$

Now suppose that $\psi \notin N^T$. Then $g(s)$ is a nonzero entire function of exponential growth less than or equal to T_1. Hence the indicator function

$$h_g(\theta) = \limsup_{r \to \infty} r^{-1} \log |g(re^{i\theta})|, \ 0 \leqslant \theta \leqslant 2\pi,$$

of g satisfies $|h_g(\theta)| \leqslant T_1$ for $0 \leqslant \theta \leqslant 2\pi$ (Markushevich [102, Theorem 9.18]). In particular $h_g(0) \geqslant -T_1$ and thus

$$\limsup_{s \to +\infty} |g(s)| \, e^{s(T_1+\varepsilon)} = \infty$$

for every $\varepsilon > 0$. On the other hand, we obtain in the case $T_1 + \varepsilon < t$ that

$$\lim_{s\to+\infty} |\det \Delta(s)\, \hat{y}(s)| e^{s(T_1+\varepsilon)}$$

$$= \lim_{s\to+\infty} \left(|\det \Delta(s)| e^{-s(t-T_1-\varepsilon)}\right)\left(|\hat{y}(s)| e^{st}\right)$$

$$= 0$$

since $y(\tau) = 0$ for $\tau \leqslant t$. This is a contradiction.

The remaining identities $N^T = N_t^T$ and $cl(R) = cl(R_{t+h})$ follow from Remark 4.1.3 (iii) and (iv) ($\Gamma = 0$). □

The main idea in the proof of Lemma 4.1.10 is due to Olbrot [119, Lemma 1] who proved the corresponding result for retarded systems in the state space $C(M = 0, \Gamma = 0)$. Moreover, note that the identity $R = R_t$, $t > nh$, is known for neutral systems of the form

$$\dot{x}(t) = A_0 x(t) + A_1 x(t-h) + A_{-1}\dot{x}(t-h) + B_0 u(t) \tag{4}$$

(Banks-Jacobs-Langenhop [5, Corollary 5.1]).

4.1.11 THEOREM. *Let* $t \geqslant T_0$ *and* $t > T_1$. *Then the following statements are equivalent.*

(i) Σ *is spectrally controllable.*

(ii) Σ *is approximately null-controllable in time* $t+h$.

(iii) Ω^T *is spectrally observable.*

(iv) Ω^T *is finally observable in time* t.

In the case $\Gamma = 0$ *the following statements are equivalent to* (i), (ii), (iii), *and* (iv).

(v) Ω *is spectrally controllable.*

(vi) Ω *is approximately null-controllable in time* t+h.

(vii) Σ^T *is spectrally observable.*

(viii) Σ^T *is finally observable in time* t.

Proof. By Theorem 4.1.7 and Lemma 4.1.10, statement (i) implies that $(S(t)\phi,0) \in cl(R) = cl(R_{t+h})$ for every $\phi \in M^p$. By Remark 4.1.9 this implies (ii). Conversely, approximate null-controllability of Σ in time t+h implies that $(S(t+h)\phi,0) \in cl(R)$ for every $\phi \in M^p$ (Remark 4.1.9) and hence spectral controllability (Theorem 4.1.7).

Clearly, (i) is equivalent to (iii), and (iii) is equivalent to $N^T = N_t^T \subset \ker S^T(t)$ (Theorem 4.1.7 and Lemma 4.1.10). By definition, this means that system Ω^T is finally observable.

The remainder of the theorem follows by analogy. □

It is an open problem whether spectral controllability of a *retarded* system implies *exact* null-controllability. This has only been proved in Jacobs-Langenhop [60] for two-dimensional systems of the form (3). We mention that the equivalence of spectral controllability and exact null-controllability has also been claimed by Marchenko [101] for retarded systems with finitely many discrete delays. However, the arguments in [101] seem incomplete.

For neutral systems, such a relation is definitely false, as the following example shows.

4.1.12 EXAMPLE Consider the NFDE (4) where

$$A_0 = \begin{bmatrix} 0 & 1 \\ 0 & 0 \end{bmatrix}, \quad A_1 = \begin{bmatrix} 0 & 0 \\ 0 & 0 \end{bmatrix}, \quad A_{-1} = \begin{bmatrix} 1 & 0 \\ 0 & 1 \end{bmatrix}, \quad B_0 = \begin{bmatrix} 0 \\ 1 \end{bmatrix}.$$

Then the matrix

$$[\Delta(\lambda), B_0] = \begin{bmatrix} \lambda - \lambda e^{-\lambda h} & -1 & 0 \\ 0 & \lambda - \lambda e^{-\lambda h} & 1 \end{bmatrix}$$

is of rank 2 for every $\lambda \in \mathbb{C}$. Hence system (4) is spectrally controllable in this case (Proposition 4.1.2).

However, since rank $A_{-1} = 2$, the semigroup $S(t)$ is bijective for every $t \geq 0$ (Proposition 3.1.15). This means that exact null-controllability is equivalent to exact controllability in the state space $W^{1,p}$. But the matrix pair (A_{-1},B_0) is not controllable and hence exact controllability fails (Jacobs-Langenhop [61, Corollary 2.1]).

4.2 APPROXIMATE CONTROLLABILITY AND STRICT OBSERVABILITY

Approximate controllability properties of neutral systems in the state space $W^{1,2}$ have been investigated by O'Connor [109], O'Connor-Tarn [110] (no input delays) and Bartosiewicz [10]. In this section we describe an alternative approach to these problems. Moreover, we present the following results which are apparently new.

1* A duality relation between approximate controllability and strict observability for NFDEs with delays in input and output.

2* The equivalence of strict observability with (a) spectral observability and (b) observability of small solutions (the dual property of completability).

3* The independence of approximate controllability and strict observability from the choice of the state space ($\Gamma = 0$).

4* A controllability criterion in terms of the system matrices for a rather general class of neutral systems, extending the results of Bartosiewicz [10] and O'Connor-Tarn [110].

In order to derive satisfactory results, we have to take into account that the maximal delays in the state and input/output variables may be of different length. Therefore we assume that

$$\eta(\tau) = \eta(-h_x),\ \mu(\tau) = \mu(-h_x),\ \tau \leqslant -h_x, \tag{5.1}$$

$$\beta(\tau) = \beta(-h_u),\ \gamma(\tau) = \gamma(-h_u),\ \tau \leqslant -h_u, \tag{5.2}$$

for some h_x, $h_u \in [0,h]$. We can also assume without loss of generality that either $h_x = h$ or $h_u = h$.

CONTROLLABILITY

If (5) is satisfied, then a solution $w(t)$, $x(t)$ of Σ is already uniquely determined by an initial condition of the form

$$w(0) = \phi^0, \; x(\tau) = \phi^1(\tau), \; -h_x \leq \tau < 0, \tag{6.1}$$

$$u(\tau) = \xi(\tau), \; -h_u \leq \tau < 0, \tag{6.2}$$

(together with the input u(t), $t \geq 0$). For system Ω it suffices to consider the initial condition

$$x(\tau) = \phi(\tau), \; -h_x \leq \tau \leq 0, \tag{7.1}$$

$$u(\tau) = \xi(\tau), \; -h_u \leq \tau < 0. \tag{7.2}$$

These facts suggest the study of approximate controllability properties of Σ and Ω in the reduced state spaces

$$M_x^p \times L_u^p = R^n \times L^p([-h_x,0];\mathbb{R}^n) \times L^p([-h_u,0];\mathbb{R}^m),$$

$$W_x^{1,p} \times L_u^p = W^{1,p}([-h_x,0];\mathbb{R}^n) \times L^p([-h_u,0];\mathbb{R}^m).$$

4.2.1 DEFINITION

(i) *System* Σ *is said to be approximately controllable if for all* $\phi \in M^p$, $\xi \in L^p$, *and* $\varepsilon > 0$ *there exists a time* $t > 0$ *and an input* $u \in L^p([0,t];\mathbb{R}^m)$ *such that the corresponding forced motion* w(t), x(t) *of* Σ *with initial condition zero satisfies*

$$\left[|\phi^0 - w(t)|^p + \int_{-h}^{0} |\phi^1(\tau) - x(t+\tau)|^p d\tau + \int_{-h_u}^{0} |\xi(\tau) - u(t+\tau)|^p d\tau\right]^{1/p} < \varepsilon$$

(ii) *System* Ω *is said to be approximately controllable if for all* $\phi \in W^{1,p}$, $\xi \in L^p$, *and* $\varepsilon > 0$ *there exists a time* $t > 0$ *and an input* $u \in L^p([0,t];\mathbb{R}^m)$ *such that the corresponding forced motion* $x(t)$ *of* Ω *with initial condition zero satisfies*

$$\left[|\phi(0) - x(t)|^p + \int_{-h}^{0} |\dot{\phi}(\tau) - \dot{x}(t+\tau)|^p d\tau + \int_{-h_u}^{0} |\xi(\tau) - u(t+\tau)|^p d\tau\right]^{1/p} < \varepsilon .$$

For simplicity of notation it is convenient to introduce the restriction operators

$$r_x : M^p \to M_x^p, \; r_x : W^{1,p} \to W_x^{1,p}, \; r_u : L^p \to L_u^p$$

in an obvious way. Then the above definition can be reformulated as follows.

4.2.2 REMARK. Approximate controllability of Σ is equivalent to

$$\mathrm{cl}(\{(r_x\phi, r_u\xi) \mid (\phi,\xi) \in R\}) = M_x^p \times L_u^p \tag{8}$$

and approximate controllability of Ω to

$$\mathrm{cl}(\{(r_x\phi, r_u\xi) \mid (\phi,\xi) \in R\}) = W_x^{1,p} \times L_u^p. \tag{9}$$

OBSERVABILITY

Let us now turn to the question how to define (strict) observability for the systems Ω^T and Σ^T.

Since h_x is the maximal length of the delays in the state variable, it seems natural to consider the solutions $x(t)$ of Ω^T on the time interval $[-h_x,\infty)$. Moreover, the output $y(t)$ of Ω^T at time t depends on the values of $x(\cdot)$ on the interval $[t-h_u,t]$. Hence $y(t)$ can be defined for $t \geq h_u-h_x$. This situation is illustrated in Figure 4 for the case $h_u < h_x$.

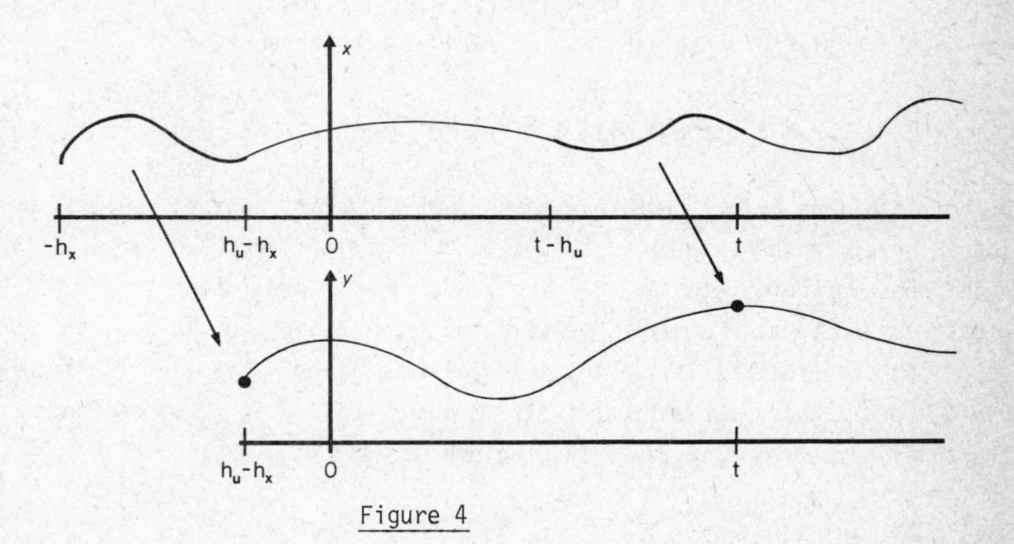

Figure 4

The above considerations suggest the following definition of strict observability.

2.3 DEFINITION. *System Ω^T is said to be strictly observable if the solutions $x(t)$ of Ω^T satisfy*

$$y(t) = 0 \;\forall\; t \geq h_u - h_x \Rightarrow x(t) = 0 \;\forall\; t \geq -h_x. \tag{10}$$

System Σ^T is said to be strictly observable if the solutions w(t), x(t) *of Σ^T satisfy*

$$y(t) = 0 \;\forall\; t \geq h_u - h_x \Rightarrow x(t) = 0 \;\forall\; t \geq -h_x. \tag{11}$$

The usual definition of (initial) observability for infinite-dimensional systems is that the initial state must be zero if the output vanishes for $t \geq 0$ (see, e.g., Curtain-Pritchard [24], Dolecki-Russell [36], Dolecki [35], Triggiani [143]). For the system Ω^T in the state space $W^{1,q}$ this is equivalent to

$$y(t) = 0 \;\forall\; t \geq 0 \Rightarrow x(t) = 0 \;\forall\; t \geq -h. \tag{12}$$

Note that (12) coincides with (10), if $h_u = h_x = h$. However, in the case $h_x < h_u = h$, observability in the sense of (12) would require the strong condition $m \geq n$ (this follows from Theorem 4.2.11 below). And if $h_u < h_x = h$, then such a notion of observability would imply the property

$$x(t) = 0 \;\forall\; t \geq -h_u \Rightarrow x(t) = 0 \;\forall\; t \geq -h.$$

for the solutions of Ω^T. But this means that Ω^T has no nonzero small solution. Therefore we restrict ourselves to the study of strict observability in the sense of Definition 4.2.3 which takes care of the length of the delays in state and output. Moreover, we will see that this notion is dual to approximate controllability in the sense of Definition 4.2.1. For this purpose we need a characterization of strict observability in terms of the structural operators G, D, $\mathcal{G}$, $\mathcal{D}$ and the embeddings

$$r_x^* : M_x^q \to M^q,\; r_x^* : W_x^{-1,q} \to W^{-1,q},\; r_u^* : L_u^q \to L^q.$$

More precisely, we make use of the (closed) range spaces of these operators.

4.2.4 REMARKS

(i) Let $\phi \in M^p$ and $g \in M^q$. Note that $r_x\phi = 0$ if and only if $\phi^0 = 0$ and

$\phi^1(\tau) = 0$ for $-h_x \leq \tau \leq 0$. Hence $g \in \operatorname{ran} r_x^* = (\ker r_x)^\perp$ iff $g^1(\tau) = 0$ for $-h \leq \tau \leq -h_x$.

(ii) Let $\xi \in L^p$ and $d \in L^q$. Note that $r_u\xi = 0$ if and only if $\xi(\tau) = 0$ for $-h_u \leq \tau \leq 0$. Hence $d \in \operatorname{ran} r_u^* = (\ker r_u)^\perp$ iff $d(\tau) = 0$ for $-h \leq \tau \leq -h_u$.

(iii) Let $\phi \in W^{1,p}$ and $g \in M^q$. Note that $r_x\phi = 0$ if and only if $\phi(\tau) = 0$ for $-h_x \leq \tau \leq 0$. We conclude that $\pi^T g \in \operatorname{ran} r_x^*$ iff $\langle \pi^T g, \phi \rangle = 0$ for every $\phi \in \ker r_x$ or equivalently

$$g^2(-t) = \int_t^h g^1(-s)ds, \; h_x \leq t \leq h.$$

4.2.5 LEMMA

(i) *System* Ω^T *is strictly observable if and only if the following implication holds for every* $g \in M^q$

$$g \in \operatorname{ran} r_x^*, \; D^*G^*g \in \operatorname{ran} r_u^*, \; G^*g \in N^T \Rightarrow g = 0. \tag{13}$$

(ii) *System* Σ^T *is strictly observable if and only if the following implication holds for every* $g \in M^q$

$$\pi^T g \in \operatorname{ran} r_x^*, \; D^*G^*\pi^T g \in \operatorname{ran} r_u^*, \; G^*\pi^T g \in N^T \Rightarrow \pi^T g = 0. \tag{14}$$

Proof. (i) Consider the system

$$\dot{z}(t) = \int_{-t}^0 d\eta^T(\tau)z(t+\tau) + \int_{-t}^0 d\mu^T(\tau)\dot{z}(t+\tau) + g^1(-t), \; t \geq 0, \tag{15.1}$$

$$z(0) = g^0 \tag{15.2}$$

$$w(t) = B^T z_t + \Gamma^T \dot{z}_t, \; t \geq h_u, \tag{15.3}$$

where $g \in \operatorname{ran} r_x^*$ which means that $g^1(-t) = 0$ for $t \geq h_x$ (see Remark 4.2.4(i)). Then every solution $z(t)$ of (15) satisfies Ω^T for $t \geq h_x$. Conversely, if $x(t)$, $t \geq -h_x$, is any solution of Ω^T, then $z(t) = x(t-h_x)$, $t \geq 0$, satisfies (15) for some $g \in \operatorname{ran} r_x^*$.

Hence Ω^T is strictly observable if and only if the solutions of (15) have the following property

$$w(t) = 0 \; \forall \; t \geq h_u \Rightarrow z(t) = 0 \; \forall \; t \geq 0. \tag{16}$$

Now recall that $G^*g = z_h \in W^{1,p}$ where $z(t)$ is the unique solution of (15.1), (15.2) corresponding to $g \in \operatorname{ran} r_x^*$ (compare Section 2.2, page 60). Consequently, the output $w(t)$, given by (15.3), vanishes for $t \geq h$ if and only if $G^*g \in N^T$. On the interval $[h_u,h]$ this output is described by

$$w(t) = \int_{-t}^{0} d\beta^T(\tau) z_h(t+\tau-h) + \int_{-t}^{0} d\gamma^T(\tau)\dot{z}_h(t+\tau-h)$$

$$= [D^*G^*g](-t),\ h_u \leq t \leq h,$$

(compare the proof of Proposition 2.3.10, page 80). Hence $w(t) = 0$ for $h_u \leq t \leq h$ if and only if $D^*G^*g \in \operatorname{ran} r_u^*$ (see Remark 4.2.4 (ii)). This shows that (16) is equivalent to (13).

(ii) Now consider the system

$$\dot{z}(t) = \int_{-t}^{0} d\eta^T(\tau)x(t+\tau) + g^1(-t),\ z(0) = g^0, \tag{17.1}$$

$$x(t) = z(t) + \int_{-t}^{0} d\mu^T(\tau)x(t+\tau) + g^2(-t),\ t \geq 0, \tag{17.2}$$

$$y(t) = B^T x_t,\ t \geq h_u, \tag{17.3}$$

where $g \in M^q$ satisfies $\pi^T g \in \operatorname{ran} r_x^*$. Then remark 2.4.2(iii) shows that every solution $x(t)$, $t \geq 0$, of (17) satisfies Σ^T for $t \geq h_x$ with $z(t)$ replaced by $z(t) + g^2(-t)$ on $[h_x,h]$. Conversely, if $x(t)$ is a solution of Σ^T, then $x(t-h_x), t \geq 0$, satisfies (17) for some $g \in M^q$ with $g^1(-t) = g^2(-t) = 0$, $h_x \leq t \leq h$.

The remainder of the proof is precisely the same as above. We obtain that Σ^T is strictly observable if and only if the solutions of (17) have the property

$$y(t) = 0\ \forall\ t \geq h_u \Rightarrow x(t) = 0\ \forall\ t \geq 0 \tag{18}$$

and that (18) is equivalent to (14). □

DUALITY

The next result shows that strict observability in the sense of Definition 4.2.3 is dual to approximate controllability in the sense of Definition 4.2.1.

4.2.6 THEOREM

(i) *System* Σ *is approximately controllable if and only if system* Ω^T *is strictly observable.*

(ii) *System* Ω *is approximately controllable if and only if system* Σ^T *is strictly observable.*

Proof. It follows from Remark 4.2.2 that system Σ is approximately controllable if and only if the following implication holds for every $g \in M^q$ and every $d \in L^q$

$$g \in \text{ran } r_x^*, \; d \in \text{ran } r_u^*, \; (g,d) \perp R \Rightarrow g = 0, \; d = 0.$$

Moreover, $(g,d) \perp R$ if and only if $G^*g \in N^T$ and $d = - D^*G^*g$ (Lemma 1.4). Hence the above implication is equivalent to (13), i.e. to the strict observability of system Ω^T (Lemma 4.2.5).

This proves (i). Statement (ii) can be established by analogy. □

The proof of the above theorem (respectively Lemma 4.2.5) is less complicated if the delays in the input are of the same length as the delays in the state ($h_u = h_x = h$) or if there are no delays in the input variable ($h_u = 0$, $h_x = h$). In the second case this duality result has been proved in Salamon [134, Theorem 3.4] for retarded systems.

OBSERVABILITY OF SMALL SOLUTIONS

Recall that system Ω^T is spectrally observable if and only if 'zero output' implies 'small solution' (Theorem 4.1.7). The remaining property for strict observability is that 'zero output' and 'small solution' imply 'zero solution' which means

$$\left.\begin{array}{l} y(t) = 0 \quad \forall\, t \geq h_u - h_x \\ x(t) = 0 \quad \forall\, t \geq h - h_x \end{array}\right\} \Rightarrow x(t) = 0 \;\forall\, t \geq -h_x. \tag{19}$$

This will be called *observability of small solutions.* The corresponding property for the solutions of system Σ^T is

$$\left.\begin{array}{l} \mathrm{y}(t) = 0 \quad \forall\, t \geq h_u - h_x \\ \mathrm{x}(t) = 0 \quad \forall\, t \geq h - h_x \end{array}\right\} \Rightarrow \mathrm{x}(t) = 0 \;\forall\, t \geq -h_x. \tag{20}$$

4.2.7 PROPOSITION

(i) *System* Ω^T *is strictly observable if and only if it is spectrally observable and satisfies* (19).

(ii) *System* Σ^T *is strictly observable if and only if it is spectrally observable and satisfies* (20).

Proof. First let Ω^T be strictly observable. Then (19) is obviously satisfied. Moreover let $\psi \in N^T$ and let $x(t)$, $t \geq -h$, be the corresponding solution of Ω^T. Then $x(t+h_x)$, $t \geq -h_x$, is a solution of Ω^T and the corresponding output $y(t+h_x)$ vanishes for $t \geq h_u-h_x$. By (10), this implies $x(t) = 0$ for $t \geq 0$. We conclude that $N^T \subset \ker S^T(h) \subset \ker S^T(T_0)$, which implies spectral observability of system Ω^T (Theorem 4.1.7).

Conversely, let Ω^T be spectrally observable and let (19) be satisfied. Moreover, let $x(t)$, $t \geq -h$, be a solution of Ω^T such that the corresponding output $y(t)$ vanishes for $t \geq h_u-h_x$. Then $S^T(h-h_x)x_0 = x_{h-h_x} \in N^T \subset \ker S^T(nh)$ (Theorem 4.1.7). Hence $x(t) = 0$ for every $t \geq nh-h_x$. By induction, it follows from (19) that $x(t) = 0$ for $t \geq -h_x$.

This proves (i). Statement (ii) can be proved by analogy. □

Clearly, the small solutions of Ω^T (respectively Σ^T) are observable if there is no nonzero small solution on the time interval $[-h_x,\infty)$. In this case strict observability is equivalent to spectral observability.

4.2.8 COROLLARY

(i) *If system* Ω^T *(respectively* Σ^T*) has no nonzero small solution, then strict observability is equivalent to spectral observability.*

(ii) *If system* Σ *(respectively* Ω*) is complete in the state space* M_x^p *(respectively* $W_x^{1,p}$*), then approximate controllability is equivalent to spectral controllability.*

INDEPENDENCE OF APPROXIMATE CONTROLLABILITY AND STRICT OBSERVABILITY FROM THE CHOICE OF THE STATE SPACE ($\Gamma = 0$)

Using the above characterization of strict observability (Proposition 4.2.7), we can prove that, in the case $\Gamma = 0$, this notion does not depend on the choice of the state space (M^q or $W^{1,q}$) for the transposed system. This means that strict observability of Ω^T is equivalent to strict observability of Σ^T.

We need the following preliminary fact.

4.2.9 LEMMA. *Let* $\Gamma = 0$ *and let* $z(t)$, $x(t)$ *be a small solution of* Σ^T *with corresponding output* $y(t)$, $t \geq h_u - h$. *Then*

$$\mathcal{x}(t) := -\int_t^{T_0} x(s)\, ds, \quad t \geq -h,$$

is a small solution of Ω^T *and satisfies*

$$B^T \mathcal{x}_t = -\int_t^{T_0} y(s)\, ds, \quad t \geq h_u - h.$$

Proof. It follows from Lemma 3.1.9 that $\mathcal{x}(t)$ is a small solution of Ω^T. Moreover

$$B^T \mathcal{x}_t = -\int_{-h_u}^{0} d\beta^T(\tau) \int_{t+\tau}^{T_0} x(s) ds = -\int_{-h_u}^{0} d\beta^T(\tau) \int_t^{T_0} x(s+\tau) ds$$

$$= -\int_t^{T_0} B^T x_s\, ds = -\int_t^{T_0} y(s)\, ds, \quad t \geq h_u - h. \quad \square$$

4.2.10 COROLLARY. *Let* $\Gamma = 0$. *Then the small solutions of* Ω^T *are observable if and only if system* Σ^T *has the same property. Moreover the following statements are equivalent.*

(i) *System* Σ *is approximately controllable.*

(ii) *System* Ω *is approximately controllable.*

(iii) *System* Ω^T *is strictly observable.*

(iv) *System* Σ^T *is strictly observable.*

Proof. It follows from Lemma 4.2.9 that (19) implies (20). The converse implication is a consequence of the fact that system Ω^T represents the restriction of system Σ^T to $W^{1,p}$-solutions (Remark 2.3.3). Hence we obtain from Proposition 4.2.7 that Ω^T is strictly observable if and only if Σ^T is. The remaining assertions of the corollary follow from the duality result (Theorem 4.2.6). $\square$

Lack of space prevents discussion of the (more or less obvious) consequences of Corollary 4.2.10 for the approximate controllability and strict observability in the state space of continuous functions (compare Corollary 3.1.12).

A MATRIX-TYPE CONDITION

Recall that Proposition 4.1.2 gives a matrix-type condition which can be checked directly in many cases. Our next result is a computable criterion for observability of small solutions.

4.2.11 THEOREM

(i) *Let* $h_x > 0$, $h_u > 0$, *and suppose that the following equations hold for some* $\varepsilon > 0$

$$\eta(\tau) = A_1 + \eta(-h_x), \ \mu(\tau) = A_{-1} + \mu(-h_x), \ -h_x < \tau < \varepsilon - h_x, \tag{21.1}$$

$$\beta(\tau) = B_1 + \beta(-h_u), \ \gamma(\tau) = B_{-1} + \gamma(-h_u), \ -h_u < \tau < \varepsilon - h_u. \tag{21.2}$$

Then the small solutions of Ω^T *are observable if and only if*

$$\text{rank } [A_1 + \lambda A_{-1}, B_1 + \lambda B_{-1}] = n \tag{22}$$

for some $\lambda \in \mathbb{C}$.

(ii) *Let* $h_x > 0$, *suppose that* (21.1) *holds for some* $\varepsilon > 0$ *and let* B *and* Γ *be given by*

$$B\xi = B_0\xi(0), \ \Gamma\xi = B_{-0}\xi(0), \ \xi \in C([-h,0];\mathbb{R}^m). \tag{23}$$

Then the small solutions of Ω^T *are observable if and only if*

$$\text{rank } [A_1 + \lambda A_{-1} \ \ B_0 + \lambda B_{-0}] = n \tag{24}$$

for some $\lambda \in \mathbb{C}$.

Proof. (i) The small solutions of Ω^T are observable if and only if the following implication holds

$$\left.\begin{array}{l} y(t) = 0 \ \forall \ t \geq h_u - h_x \\ x(t) = 0 \ \forall \ t \geq \varepsilon - h_x \end{array}\right\} \Rightarrow x(t) = 0 \ \forall \ t \geq -h_x. \tag{25}$$

Now let (21) be satisfied and define $x(t) := x(t-h_x)$, $f(t) := \dot{x}(t-h_x)$ for $0 \leq t \leq \varepsilon$. Then (25) is equivalent to

$$\left.\begin{array}{l}\dot{x}(t) = f(t),\ x(\varepsilon) = 0 \\ 0 \equiv A_1^T x(t) + A_{-1}^T f(t) \\ 0 \equiv B_1^T x(t) + B_{-1}^T f(t)\end{array}\right\} \Rightarrow x(t) \equiv 0. \tag{26}$$

This means that

$$\operatorname{rank}\begin{bmatrix}\lambda I & -I \\ A_1^T & A_{-1}^T \\ B_1^T & B_{-1}^T\end{bmatrix} = n + \operatorname{rank}\begin{bmatrix}-I \\ A_{-1}^T \\ B_{-1}^T\end{bmatrix} = 2n$$

for some $\lambda \in \mathbb{C}$ (see Appendix, Theorem A6). This rank condition is equivalent to (22).

(ii) Under the assumptions of (ii), we obtain that (25) is equivalent to (26) with B_1 and B_{-1} replaced by B_0 and B_{-0}. □

The above criterion was first obtained by Manitius and Triggiani [98] as a necessary condition for approximate controllability of a retarded system of the form (3) (a single point delay, no input delays) in the state space M^2. In this case (24) reduces to

$$\operatorname{rank}\,[A_1, B_0] = n. \tag{27}$$

Manitius [94] has generalized condition (27) to the case of finitely many point delays. Moreover it has been shown in [94] that (27) together with spectral controllability is also sufficient for approximate controllability of a RFDE in the state space M^2.

For neutral systems an analogous result in the state space $W^{1,2}$ has been derived independently by Bartosiewicz [10] and O'Connor-Tarn [110]. They have shown that system (4) is approximately controllable in the state space $W^{1,2}$ if and only if it is spectrally controllable and

$$\operatorname{rank}\,[A_1 + \lambda A_{-1}, B_0] = n \tag{28}$$

for some $\lambda \in \mathbb{C}$. Our results generalize their criterion to the fairly general situation of Theorem 4.2.11. More precisely, we obtain the following characterization of approximate controllability as a consequence of Theorem 4.2.6, Proposition 4.2.7 and Theorem 4.2.11.

4.2.12 CORLLARY

(i) *Let* $h_x > 0$, $h_u > 0$, *and suppose that* (21) *holds for some* $\varepsilon > 0$. *Then system* Σ *is approximately controllable if and only if it is spectrally controllable and* (22) *holds for some* $\lambda \in \mathbb{C}$.

(ii) *Let* $h_x > 0$, $h_u = 0$, *suppose that* (21.1) *holds for some* $\varepsilon > 0$, *and let* B *and* Γ *be given by* (23). *Then system* Σ *is approximately controllable if and only if it is spectrally controllable and* (24) *holds for some* $\lambda \in \mathbb{C}$.

The trivial example of an uncontrolled, complete system shows that spectral controllability may fail while condition (22), respectively (24), is satisfied. The reverse situation may also occur.

4.2.13 EXAMPLES

(i) The scalar n-th order differential-difference equation

$$z^{(n)}(t) = \sum_{j=0}^{n-1} \alpha_j z^{(j)}(t) + \sum_{j=0}^{n} \beta_j z^{(j)}(t-h) + u(t) \tag{29}$$

can be rewritten as a system of the type (4) where the matrices A_0, A_1, A_{-1}, B_0 are given by (3.18) and

$$B_0 = [0 \; \dots \; 0 \; 1]^T. \tag{30}$$

It is easy to see that this system is spectrally controllable; but rank $[A_1 + \lambda A_{-1} \; B_0] = 1$ and hence (24) fails unless $n = 1$.

(ii) The two-dimensional system

$$\begin{aligned} \dot{x}_1(t) &= x_1(t-h) + \dot{x}_2(t-h) \\ \dot{x}_2(t) &= -x_1(t) + u(t) \end{aligned} \tag{31}$$

is described by the matrices

$$A_0 = \begin{bmatrix} 0 & 0 \\ -1 & 0 \end{bmatrix}, \quad A_1 = \begin{bmatrix} 1 & 0 \\ 0 & 0 \end{bmatrix}, \quad A_{-1} = \begin{bmatrix} 0 & 1 \\ 0 & 0 \end{bmatrix}, \quad B_0 = \begin{bmatrix} 0 \\ 1 \end{bmatrix}. \tag{32}$$

The matrix

$$\left[\Delta(\lambda), B_0\right] = \begin{bmatrix} \lambda - e^{-\lambda h} & -\lambda e^{-\lambda h} & 0 \\ 1 & \lambda & 1 \end{bmatrix}$$

is of rank 2 for every $\lambda \in \mathbb{C}$, and condition (24) is obviously satisfied. Hence system (31) is approximately controllable.

(iii) The slightly modified system

$$\begin{aligned} \dot{x}_1(t) &= x_1(t-h) + \dot{x}_2(t-h) + u(t-h) \\ \dot{x}_2(t) &= -x_1(t) \end{aligned} \tag{33}$$

can be written in the form

$$d/dt\ (x(t) - A_{-1}x(t-h)) = A_0x(t) + A_1x(t-h) + B_1u(t-h) \tag{34}$$

where A_0, A_1, A_{-1} are as in (ii) and

$$B_1 = \begin{bmatrix} 1 \\ 0 \end{bmatrix}. \tag{35}$$

A glance at the matrix

$$\left[\Delta(\lambda),\ B(e^{\lambda\cdot})\right] = \begin{bmatrix} \lambda - e^{-\lambda h} & -\ \lambda e^{-\lambda h} & e^{-\lambda h} \\ 1 & \lambda & 0 \end{bmatrix} \tag{36}$$

shows that (33) is still spectrally controllable; but this time (22) fails.

COMPLETABILITY

We have seen that strict observability of system Ω^T, respectively Σ^T, splits up into two (independent) properties, namely spectral observability and observability of small solutions (Proposition 4.2.7). By duality (Theorem 4.2.6), approximate controllability of Σ, respectively Ω, is equivalent to the same two properties. Moreover, spectral observability is clearly dual to spectral controllability. But what is the systems theoretic meaning of 'observability of small solutions' for the control systems Σ and Ω?

A special answer can be given from Chapter 3. If Σ is complete, then Ω^T has no nonzero small solution (Theorem 3.1.10) and hence (19) is satisfied. In this case approximate controllability is equivalent to spectral controllability (Corollary 4.2.8).

A more satisfactory answer can be given if we assume that $h_u \leqslant h_x = h$. In this case we will show (under the assumptions of Theorem 4.2.11) that the observability of the small solutions of system Ω^T is equivalent to the existence of a feedback

$$u(t) = Kx(t-h+h_u) \tag{37}$$

($K \in \mathbb{R}^{m \times n}$) such that the closed loop system Σ,(37) is complete. This property is called *completability*. It plays a central role in the theory of Manitius [94, 95] and Bartosiewicz [10]. This is the reason why Bartosiewicz [10] was able to treat only the case $h_u \leqslant h_x$.

4.2.14 <u>LEMMA</u>. *Let* $h_u \leqslant h_x = h$.

(i) *If system* Σ *is completable, then the small solutions of system* Ω^T *are observable.*

(ii) *Let* $h > 0$ *and let* (21.1) *be satisfied for some* $\varepsilon > 0$. *Moreover, suppose that either* $h_u = 0$ *or* (21.2) *holds. Then* Σ *is completable if and only if the small solutions of* Ω^T *are observable.*

<u>Proof</u>. (i) Let $K \in \mathbb{R}^{m \times n}$ such that the system Σ,(37) is complete. Moreover let $x(t)$ be a solution of Ω^T which vanishes for $t \geqslant 0$ and whose output $y(t) = B^T x_t + \Gamma^T \dot{x}_t$ vanishes for $t \geqslant h_u - h$. Then, for $t \geqslant 0$,

$$\begin{aligned}\dot{x}(t) &= L^T x_t + M^T \dot{x}_t + K^T y(t-h+h_u) \\ &= L^T x_t + K^T B^T x_{t-h+h_u} + M^T \dot{x}_t + K^T \Gamma^T \dot{x}_{t-h+h_u}.\end{aligned}$$

But this is precisely the transposed of the complete system Σ,(37). Hence $x(t) = 0$ for $t \geqslant -h$.

(ii) First let $h_u > 0$. Then the small solutions of Ω^T are observable if and only if (22) is satisfied (Theorem 4.2.11). It has been proved by Manitius [94, Lemma 12] that (22) implies the existence of a real $n \times m$-matrix K (with entries 0 or 1) such that rank $(A_1 + \lambda A_{-1} + KB_1 + \lambda KB_{-1}) = n$. By Theorem 3.1.10, this means that the closed loop system Σ,(37) is complete.

The case $h_u = 0$ can be treated by analogy. □

It is an open question whether the assumptions in Statement (ii) of the above lemma are really necessary to prove the equivalence of completability (closed loop property) and observability of small solutions (open loop property) for the transposed system.

AN OPERATOR-TYPE CONDITION

In two special cases ($h_u = 0$, $h_u = h_x$) we characterize the observability of small solutions in terms of the structural operators. For this purpose we need some preliminary facts.

4.2.15 REMARKS

(i) Let $\psi \in W^{1,q}$ and let $x(t)$, $t \geq -h$, be the corresponding solution of Ω^T with output $y(t)$. Then, by Proposition 2.3.10,

$$F^*\psi = 0,\ E^*\psi = 0 \iff x(t) = 0,\ y(t) = 0\ \forall\ t \geq 0.$$

Moreover, the following equation holds for $h_u - h \leq t \leq 0$ (equivalently $-h \leq -t-h \leq -h_u$)

$$y(t) = \int_{-t-h}^{0} d\beta^T(\tau)x(t+\tau) + \int_{-t-h}^{0} d\gamma^T(\tau)\dot{x}(t+\tau) = [D^*\psi](-t-h).$$

(ii) Let $\psi \in M^q$ and let $z(t)$, $x(t)$ be the corresponding solution of Σ^T with output $y(t)$. Then, by Proposition 2.3.10,

$$\mathcal{F}^*\psi = 0,\ \mathcal{E}^*\psi = 0 \iff x(t) = 0,\ y(t) = 0\ \forall\ t \geq 0.$$

Moreover, the following equation holds for $h_u - h \leq t \leq 0$ (equivalently $-h \leq -t-h \leq -h_u$)

$$y(t) = \int_{-t-h}^{0} d\beta^T(\tau)x(t+\tau) = [\mathcal{D}^*\psi](-t-h).$$

(iii) Let B and Γ be given by (23) and $B_{-0} = 0$. Then the operator $\mathcal{D} : L^p \to M^p$ maps $\xi \in L^p$ into the pair $\mathcal{D}\xi = (0, B_0\xi(-h-.)) \in M^p$ and $D = \iota^{T*}\mathcal{D} : L^p \to W^{-1,p}$ (Lemma 2.3.11). Moreover, in this case, $E = 0$ and $\mathcal{E} = 0$.

4.2.16 COROLLARY *Let $h_u = h_x = h$. Then*

(i) *the small solutions of Ω^T are observable iff*

$$\ker F^* \cap \ker E^* = \{0\}, \tag{38}$$

(ii) *the small solutions of Σ^T are observable iff*

$$\ker F^* \cap \ker \mathcal{E}^* = \{0\}. \tag{39}$$

4.2.17 <u>COROLLARY</u>. *Let* $h_u = 0$ *and* $h_x = h$. *Then*

(i) *the small solutions of* Ω^T *are observable iff*

$$\ker F^* \cap \ker D^* = \{0\}, \tag{40}$$

(ii) *the small solutions of* Σ^T *are observable iff*

$$\ker F^* \cap \ker \mathcal{D}^* = \{0\}. \tag{41}$$

Note that (41) is precisely the controllability condition which has been obtained by Manitius [94] for retarded systems.

4.3 F-CONTROLLABILITY AND OBSERVABILITY

Arguments analogous to those at the beginning of Section 3.2 indicate that approximate controllability in the sense of Definition 4.2.1 may be too restrictive a property for a large class of systems. In particular, the scalar n-th order differential-difference equation (29) is not approximately controllable, but has very nice properties form a control point of view, namely it is exactly null-controllable, feedback stabilizable and spectrally controllable.

The dual observability concept of approximate controllability (strict observability in the sense of Definition 4.2.3) is concerned with the past values of the solutions of Ω^T and Σ^T. However, in many cases it may be enough to have an information on the solution at times $t \geq 0$. A corresponding observability notion has been investigated for retarded systems, for instance, by Olbrot [115, 116, 119], Lee [82], Lee-Olbrot [83] and Kwong [81].

In [91] and [95] Manitius has introduced the weaker concept of approximate F-controllability for retarded systems with undelayed input variables in the product space M^2. This notion has something to do with the dual state concept (forcing terms). The idea of F-controllability has also been applied to neutral systems in the state space $W^{1,2}$ by O'Connor [109]; however, in [109] there are no further results in this direction.

In this section we develop the concept of F-controllability for NFDEs with input delays in the state spaces M^p and $W^{1,p}$. For the new controllability concept it is no longer necessary to take care of different lengths of maximal delays in state and input. This job is done automatically by the operators F and E, which allows a more elegant presentation of the results than was possible in Section 4.2.

Recall that the reachable subspace of system $\tilde{\Sigma}$ (dual state concept) is given by

$$[F\ E]R = \{F\phi + E\xi \mid (\phi,\xi) \in R\} \subset W^{-1,p}$$

(Remark 4.1.3). A suitable candidate for the closure of this subspace may be the closure of ran F + ran E in $W^{-1,p}$. Correspondingly, the closure of the reachable subspace

$$[\mathcal{F}\ \mathcal{E}]\mathcal{R} = \{\mathcal{F}\phi + \mathcal{E}\xi \mid (\phi,\xi) \in \mathcal{R}\} \subset M^p$$

of system $\tilde{\Omega}$ may be regarded in the closure of ran $\mathcal{F}$ + ran $\mathcal{E}$ in M^p. This suggests the following concept of approximate F-controllability.

4.3.1 DEFINITION. *System Σ is said to be (approximately)* F-*controllable if* cl([F E]R) = cl(ran [F E]).

System Ω is said to be (approximately) F-*controllable if* cl([$\mathcal{F}$ $\mathcal{E}$]$\mathcal{R}$) = cl(ran [$\mathcal{F}$ $\mathcal{E}$]).

Let us first check that F-controllability in the above sense is in fact a weaker property than approximate controllability in the sense of Definition 4.2.1. For this purpose, assume that (5) is satisfied and let $\phi \in M^p$, $\xi \in L^p$ be given. Then $F\phi \in W^{-1,p}$ depends only on the values $\phi(\tau)$ for $-h_x \leq \tau \leq 0$ (i.e. on $r_x\phi$) and $E\xi \in W^{-1,p}$ depends only on the values $\xi(\tau)$ for $-h_u \leq \tau \leq 0$ (i.e. on $r_u\xi$). Hence approximate controllability of Σ implies approximate F-controllability.

As has been indicated above, we will also introduce a weaker notion of observability which is only related to the values of the solution of the respective equation at times $t \geq 0$.

4.3.2 DEFINITION. *System Ω^T is said to be observable if the solutions $x(t)$ of Ω^T satisfy*

$$y(t) = 0 \ \forall\ t \geq 0 \Rightarrow x(t) = 0 \ \forall\ t \geq 0. \tag{42}$$

System Σ^T is said to be observable if the solution pairs z(t), x(t) *of* Σ^T

satisfy

$$y(t) = 0 \ \forall \ t \geqslant 0 \Rightarrow x(t) = 0 \ \forall \ t \geqslant 0. \tag{43}$$

It follows from a little time shift that this type of observability is weaker than strict observability in the sense of Definition 4.2.3. In fact, let system Ω^T be strictly observable and let $x(t)$, $t \geqslant -h$, be a solution of Ω^T with zero output for $t \geqslant 0$. Then $x(t+h_x)$, $t \geqslant -h_x$, is a solution of Ω^T with zero output for $t \geqslant h_u-h_x$. Hence $x(t) = 0$ for $t \geqslant 0$.

By Remark 4.2.15, we can reformulate Definition 4.3.2 in terms of the structural operators.

4.3.3 REMARKS

(i) System Ω^T is observable if and only if $N^T \subset \ker F^*$ or equivalently $N^T = \ker F^* \cap \ker E^*$.

(ii) System Σ^T is observable if and only if $N^T \subset \ker F^*$ or equivalently $N^T = \ker F^* \cap \ker E^*$.

OBSERVABILITY OF NONTRIVIAL SMALL SOLUTIONS

Clearly, the existence of (nonzero) trivial small solutions (Definition 3.2.2) with zero output for $t \geqslant 0$ does not affect observability in the sense of Definition 4.3.2. What is needed for this type of observability is that 'zero output' and 'small solution' imply 'trivial small solution'. This property will be called *observability of nontrivial small solutions*. For the system Ω^T this is equivalent to

$$\left.\begin{array}{l} y(t) = 0 \ \forall \ t \geqslant 0 \\ x(t) = 0 \ \forall \ t \geqslant h \end{array}\right\} \Rightarrow \quad x(t) = 0 \ \forall \ t \geqslant 0, \tag{44}$$

and for the system Σ^T to

$$\left.\begin{array}{l} y(t) = 0 \ \forall \ t \geqslant 0 \\ x(t) = 0 \ \forall \ t \geqslant h \end{array}\right\} \Rightarrow \quad x(t) = 0 \ \forall \ t \geqslant 0. \tag{45}$$

Making use of Remark 4.2.15 and Proposition 2.3.10, we can reformulate these implications in terms of structural operators.

4.3.4 REMARKS

(i) The nontrivial small solutions of Ω^T are observable if and only if

$$\ker F^*G^*F^* \cap \ker E^*G^*F^* \cap \ker [D^*G^*F^* + E^*] \subset \ker F^*. \tag{46}$$

(ii) The nontrivial small solutions of Σ^T are observable if and only if

$$\ker \mathcal{F}^*\mathcal{G}^*\mathcal{F}^* \cap \ker \mathcal{E}^*\mathcal{G}^*\mathcal{F}^* \cap \ker [\mathcal{D}^*\mathcal{G}^*\mathcal{F}^* + \mathcal{E}^*] \subset \ker \mathcal{F}^*. \tag{47}$$

Let us have a look at the special case of a system with undelayed input variables. In this case $\mathcal{E} = 0$ and the operator $\mathcal{D}$ is of the special form described in Remark 4.2.15 (iii). Hence condition (47) reduces to

$$\ker \mathcal{F}^*\mathcal{G}^* \cap \ker \mathcal{D}^*\mathcal{G}^* \cap \operatorname{ran} \mathcal{F}^* = \{0\}. \tag{48}$$

This is precisely the necessary condition for F-controllability which has been obtained by Manitius [95] for retarded systems. In Salamon [134] it has been proved that this condition, together with spectral controllability, is also sufficient.

Our main results in the general case are summarized in the theorem below. In particular, we prove that F-controllability is dual to observability and that the latter is equivalent to spectral observability and observability of small solutions.

4.3.5 THEOREM. *The following statements are equivalent.*

(i) *System Σ is* F-*controllable.*

(ii) *System Σ is spectrally controllable and*

$$\operatorname{cl}(\operatorname{ran} FGF + \operatorname{ran} FGE + \operatorname{ran} [FGD + E]) = \operatorname{cl}(\operatorname{ran} [F\ E]). \tag{49}$$

(iii) *System Ω^T is observable.*

(iv) *System Ω^T is spectrally observable and the nontrivial small solutions of Ω^T are observable.*

In the case $\Gamma = 0$ the following statements are equivalent to (i), (ii), (iii) *and* (iv).

(v) *System Ω is $\mathcal{F}$-controllable*

(vi) *System Ω is spectrally controllable and*

$$\mathrm{cl}(\mathrm{ran}\, FGF + \mathrm{ran}\, FGE + \mathrm{ran}\, [FGD + E]) = \mathrm{cl}(\mathrm{ran}\, [F\ E]). \tag{50}$$

(vii) *System Σ^T is observable.*

(viii) *System Σ^T is spectrally observable and the nontrivial small solutions of Σ^T are observable.*

Proof. '(i) $\Leftrightarrow$ (iii)' System Σ is F-controllable if and only if the following implication holds for every $\psi \in W^{1,q}$

$$(F^*\psi, E^*\psi) \perp R \Rightarrow F^*\psi = 0,\ E^*\psi = 0.$$

By Lemma 4.1.4, this is equivalent to $N^T \subset \ker F^* \cap \ker E^*$ and hence to observability of Ω^T (Remark 4.3.3 (i)).

'(iii) $\Leftrightarrow$ (iv)' First let Ω^T be observable. Then we have $N^T \subset \ker F^* = \ker S^T(h) \subset \ker S^T(T_0)$ and hence Ω^T is spectrally observable (Theorem 4.1.7). Observability of the nontrivial small solutions is a trivial consequence of observability.

Conversely, let (iv) be satisfied and let $x(t)$, $t \geq -h$, be a solution of Ω^T with zero output for $t \geq 0$. Then it follows from Theorem 4.1.7 that $x_0 \in \ker S^T(T_0) \subset \ker S^T(nh)$ and hence $x(t) = 0$ for $t \geq (n-1)h$. Now we make use of the fact that the nontrivial small solutions of Ω^T are observable, and obtain by induction that $x(t) = 0$ for $t \geq 0$. Hence Ω^T is observable.

'(iv) $\Leftrightarrow$ (ii)' This equivalence follows immediately from Remark 4.3.4 (i).

The equivalence of (v), (vi), (vii) and (viii) can be proved by analogy. Hence it remains to show that (iv) is equivalent to (viii) if $\Gamma = 0$. But the observability of Ω^T follows from that of Σ^T since, in the case $\Gamma = 0$, system Ω^T is the restriction of Σ^T to $W^{1,p}$-solutions (Remark 2.3.3). Conversely, it follows from Lemma 4.2.9 that the observability of the small solutions of Ω^T implies the same property for system Σ^T. □

Clearly, the nontrivial small solutions of Ω^T (respectively Σ^T) are observable if there is no nontrivial small solution. In this case observability is equivalent to spectral observability.

4.3.6 COROLLARY

(i) *If system Ω^T (respectively Σ^T) has no nontrivial small solution, then observability is equivalent to spectral observability.*

(ii) *If system* Σ *(respectively* Ω*) is* F-*complete, then* F-*controllability is equivalent to spectral controllability.*

A MATRIX-TYPE CONDITION

Our next result is a computable criterion for observability and F-controllability in the case of systems with a single point delay. This means that L, M, B, and Γ are given by

$$L\phi = A_0\phi(0) + A_1\phi(-h), \quad \phi \in C, \tag{51.1}$$

$$M\phi = A_{-1}\phi(-h), \quad \phi \in C, \tag{51.2}$$

$$B\xi = B_0\xi(0) + B_1\xi(-h), \quad \xi \in C([-h,0];\mathbb{R}^m), \tag{51.3}$$

$$\Gamma\xi = B_{-0}\xi(0) + B_{-1}\xi(-h), \quad \xi \in C([-h,0];\mathbb{R}^m). \tag{51.4}$$

4.3.7 THEOREM. *Let* L, M, B, Γ *be given by* (51). *Then the nontrivial small solutions of* Ω^T *are observable if and only if*

$$\max_{\lambda\in\mathbb{C}} \operatorname{rank} \begin{bmatrix} A_0-\lambda I & A_1+\lambda A_{-1} & B_0+\lambda B_{-0} & B_1+\lambda B_{-1} \\ A_1+\lambda A_{-1} & 0 & B_1+\lambda B_{-1} & 0 \end{bmatrix}$$

$$= n + \max_{\lambda\in\mathbb{C}} \operatorname{rank}\,[A_1+\lambda A_{-1} \quad B_1+\lambda B_{-1}]. \tag{52}$$

Proof. Let $K \in \mathbb{N}$ be the maximal rank of the matrix

$$[A(\lambda)\ B(\lambda)] = \begin{bmatrix} A_0-\lambda I & A_1+\lambda A_{-1} & B_0+\lambda B_{-0} & B_1+\lambda B_{-1} \\ A_1+\lambda A_{-1} & 0 & B_1+\lambda B_{-1} & 0 \end{bmatrix}$$

and $k \in \mathbb{N}$ the maximal rank of $[A_1+\lambda A_{-1}\ B_1+\lambda B_{-1}]$. Then K is always less than or equal to n + k.

NECESSITY

Suppose that $K < n + k$. Then we prove in three steps that the nontrivial small solutions of Ω^T are not observable.

Step 1 There exist polynomials

$$p(\lambda) = \sum_{j=0}^{\ell} p_j\lambda^j, \ q(\lambda) = \sum_{j=0}^{\ell} q_j\lambda^j$$

in $\mathbb{R}^n[\lambda]$ such that $p(\lambda) \not\equiv 0$ and

$$\left(p^T(\lambda)\ q^T(\lambda)\right)\left[A(\lambda)\ B(\lambda)\right] = 0 \quad \forall\ \lambda \in \mathbb{C}. \tag{53}$$

Proof. Let $M(\lambda)$ and $N(\lambda)$ be unimodular matrices of appropriate size such that

$$M(\lambda)\ [A(\lambda)\ B(\lambda)]\ N(\lambda) = \begin{vmatrix} \alpha_1(\lambda) & & & 0 & \cdot\ \cdot & 0 \\ & \ddots & & \vdots & & \vdots \\ & & \alpha_K(\lambda) & 0 & \cdot\ \cdot & 0 \\ 0 & \cdots & 0 & 0 & \cdot\ \cdot & 0 \\ \vdots & & \vdots & \vdots & & \vdots \\ 0 & \cdots & 0 & 0 & \cdot\ \cdot & 0 \end{vmatrix}$$

is in Smith form. Then the last $2n - K$ rows $\left({p^j}^T(\lambda)\ {q^j}^T(\lambda)\right)$, $j = K+1,\ldots,2n$, of $M(\lambda)$ satisfy (53). Now suppose that the polynomial vectors $p^j(\lambda)$ vanish identically. Then the $q^j(\lambda)$ are linearly independent (for every $\lambda \in \mathbb{C}$) and satisfy

$${q^j}^T(\lambda)\ [A_1+\lambda A_{-1}\ \ B_1+\lambda B_{-1}] \equiv 0.$$

This implies that the maximal rank of the matrix $[A_1+\lambda A_{-1}\ \ B_1+\lambda B_{-1}]$ is less than or equal to $n - (2n - K) = K - n < k$, a contradiction. □

Step 2

$$A_0^Tp_0 + A_1^Tq_0 = 0,\ p_\ell = A_{-1}^Tq_\ell,$$
$$p_j = A_0^Tp_{j+1} + A_1^Tq_{j+1} + A_{-1}^Tq_j,\ j = 0,\ldots,\ell-1. \tag{54.1}$$

$$A_1^Tp_0 = 0,\ A_{-1}^Tp_\ell = 0,$$
$$A_1^Tp_{j+1} + A_{-1}^Tp_j = 0,\ j = 0,\ldots,\ell-1. \tag{54.2}$$

$$B_0^Tp_0 + B_1^Tq_0 = 0,\ B_{-0}^Tp_\ell + B_{-1}^Tq_\ell = 0,$$
$$B_0^Tp_{j+1} + B_{-0}^Tp_j + B_1^Tq_{j+1} + B_{-1}^Tq_j = 0,\ j = 0,\ldots,\ell-1. \tag{55.1}$$

$$B_1^T p_0 = 0, \; B_{-1}^T p_\ell = 0,$$
$$B_1^T p_{j+1} + B_{-1}^T p_j = 0, \; j = 0,\ldots,\ell-1. \tag{55.2}$$

<u>Proof</u>. These equations follow from (53) by comparison of the coefficients. In particular, the following equation holds

$$\begin{aligned} 0 &= [A_0^T - \lambda I]p(\lambda) + [A_1^T + \lambda A_{-1}^T]q(\lambda) \\ &= \sum_{j=0}^{\ell} (A_0^T p_j + A_1^T q_j)\lambda^j - \sum_{j=0}^{\ell} (p_j - A_{-1}^T q_j)\lambda^{j+1} \\ &= A_0^T p_0 + A_1^T q_0 - (p_\ell - A_{-1}^T q_\ell)\lambda^{\ell+1} \\ &\quad + \sum_{j=1}^{\ell} (A_0^T p_j + A_1^T q_j + A_{-1}^T q_{j-1} - p_{j-1})\lambda^j. \end{aligned}$$

This proves (54.1). Equations (54.2) and (55) can be established by analogy.□

<u>Step 3</u> Let $x(t)$, $t \geqslant -h$, be defined by

$$x(t) = \begin{cases} \sum_{j=1}^{\ell+1} \left(q_{\ell+1-j} \frac{t^j}{j!} + p_{\ell+1-j} \frac{(t-h)^j}{j!} \right), & -h \leqslant t < 0, \\ \sum_{j=1}^{\ell+1} p_{\ell+1-j} \frac{(t-h)^j}{j!}, & 0 \leqslant t < h, \\ 0, & h \leqslant t < \infty. \end{cases}$$

Then $x(t)$ is absolutely continuous for $t \geqslant 0$, not identically zero on $[0,h]$, and satisfies the equations

$$\dot{x}(t) = A_0^T x(t) + A_1^T x(t-h) + A_{-1}^T \dot{x}(t-h), \tag{56}$$

$$y(t) = B_0^T x(t) + B_1^T x(t-h) + B_{-0}^T \dot{x}(t) + B_{-1}^T \dot{x}(t-h) = 0, \tag{57}$$

for $t \geqslant 0$.

Proof. These equations can be proved straightforwardly by the use of (54) and (55). We will only show that (56) holds for $0 \leq t \leq h$.

$$\begin{aligned}
\dot{x}(t) &= \sum_{j=0}^{\ell} p_{\ell-j} \frac{(t-h)^j}{j!} \\
&= \sum_{j=1}^{\ell} \left(A_0^T p_{\ell+1-j} + A_1^T q_{\ell+1-j}\right) \frac{(t-h)^j}{j!} + \sum_{j=0}^{\ell} A_{-1}^T q_{\ell-j} \frac{(t-h)^j}{j!} \\
&= A_0^T x(t) + A_1^T \sum_{j=1}^{\ell+1} q_{\ell+1-j} \frac{(t-h)^j}{j!} + A_{-1}^T \sum_{j=0}^{\ell} q_{\ell-j} \frac{(t-h)^j}{j!} \\
&\quad + \sum_{j=1}^{\ell+1} A_1^T p_{\ell+1-j} \frac{(t-2h)^j}{j!} + \sum_{j=0}^{\ell} A_{-1}^T p_{\ell-j} \frac{(t-2h)^j}{j!} \\
&= A_0^T x(t) + A_1^T x(t-h) + A_{-1}^T \dot{x}(t-h), \quad 0 \leq t \leq h. \quad \square
\end{aligned}$$

Step 3 shows that the nontrivial small solutions of system Ω^T are not observable.

SUFFICIENCY

Suppose that $K = n + k$ and let $x(t)$, $t \geq -h$, be a solution of Ω^T such that $x(t) = 0$ for $t \geq h$ and the corresponding output $y(t)$ vanishes for $t \geq 0$. Then we prove in four steps that $x(t) = 0$ for $t \geq 0$.

Step 1 The complex functions

$$\hat{x}(\lambda) = \int_0^h e^{-\lambda t} x(t)dt, \; \hat{\hat{x}}(\lambda) = \int_0^{2h} e^{-\lambda t} x(t-h)dt$$

satisfy the equation

$$\begin{aligned}
&\left(\hat{x}^T(\lambda), \hat{\hat{x}}^T(\lambda)\right) \begin{bmatrix} A_0-\lambda I & A_1+\lambda A_{-1} & B_0+\lambda B_{-0} & B_1+\lambda B_{-1} \\ A_1+\lambda A_{-1} & 0 & B_1+\lambda B_{-1} & 0 \end{bmatrix} \\
&= \left(x^T(0), x^T(-h)\right) \begin{bmatrix} -I & A_{-1} & B_{-0} & B_{-1} \\ A_{-1} & 0 & B_{-1} & 0 \end{bmatrix} =: x^T.
\end{aligned} \tag{58}$$

Proof. For every $\lambda \in \mathbb{C}$, we have

$$[A_1^T + \lambda A_{-1}^T]\hat{x}(\lambda)$$

$$= \int_0^h e^{-\lambda t} A_1^T x(t)dt + \int_0^h \lambda e^{-\lambda t} A_{-1}^T x(t)dt$$

$$= \int_0^h e^{-\lambda t}\Big(A_1^T x(t) + A_{-1}^T \dot{x}(t)\Big)dt + A_{-1}^T x(0)$$

$$= \int_0^h e^{-\lambda t}\Big(\dot{x}(t+h) - A_0^T x(t+h)\Big)dt + A_{-1}^T x(0)$$

$$= A_{-1}^T x(0),$$

$$[A_0^T - \lambda I]\hat{x}(\lambda) + [A_1^T + \lambda A_{-1}^T]\hat{\hat{x}}(\lambda)$$

$$= \int_0^h e^{-\lambda t}\Big(A_0^T x(t) + A_1^T x(t-h)\Big)dt + \int_0^h \lambda e^{-\lambda t}\Big(A_{-1}^T x(t-h) - x(t)\Big)dt$$

$$+ \Big[A_1^T + \lambda A_{-1}^T\Big] \int_h^{2h} e^{-\lambda t} x(t-h)dt$$

$$= \int_0^h e^{-\lambda t}\Big(A_0^T x(t) + A_1^T x(t-h) + A_{-1}^T \dot{x}(t-h) - \dot{x}(t)\Big)dt$$

$$- e^{-\lambda h} A_{-1}^T x(0) + A_{-1}^T x(-h) - x(0) + [A_1^T + \lambda A_{-1}^T] e^{-\lambda h}\hat{x}(\lambda)$$

$$= A_{-1}^T x(-h) - x(0).$$

The other two equations can be proved by analogy. □

Step 2 There exist matrices $A_1(\lambda) \in \mathbb{R}^{k \times n}[\lambda]$, $B_1(\lambda) \in \mathbb{R}^{k \times n}[\lambda]$, such that

$$\max_{\lambda \in \mathbb{C}} \text{rank} \begin{bmatrix} A_0-\lambda I & A_1+\lambda A_{-1} & B_0+\lambda B_{-0} & B_1+\lambda B_{-1} \\ A_1(\lambda) & 0 & B_1(\lambda) & 0 \end{bmatrix} = n + k, \tag{59}$$

$$\max_{\lambda \in \mathbb{C}} \text{rank}\, [A_1(\lambda), B_1(\lambda)] = k, \tag{60}$$

and for almost every $\lambda \in \mathbb{C}$

$$\ker\, [A_1(\lambda), B_1(\lambda)] = \ker\, [A_1+\lambda A_{-1}, B_1+\lambda B_{-1}]. \tag{61}$$

Proof. By assumption, rank $[A(\lambda_0), B(\lambda_0)] = n+k$ for some $\lambda_0 \in \mathbb{C}$. Hence the matrix $[A(\lambda_0), B(\lambda_0)]$ has $n + k$ linearly independent rows. Precisely k of these are contained in the lower $n \times (2n+2m)$-block of the matrix $[A(\lambda_0), B(\lambda_0)]$ which is given by $[A_1+\lambda_0 A_{-1}, 0, B_1+\lambda_0 B_{-1}, 0]$. Now let the matrices $A_1(\lambda)$, $B_1(\lambda)$ consist of the corresponding k rows of the matrices $A_1+\lambda A_{-1}$, $B_1+\lambda B_{-1}$. Then $A_1(\lambda)$ and $B_1(\lambda)$ have the desired properties. □

Step 3 There exists a rational matrix $T(\lambda) \in \mathbb{R}^{n \times k}(\lambda)$ such that

$$A_1+\lambda A_{-1} = T(\lambda)A_1(\lambda), \quad B_1+\lambda B_{-1} = T(\lambda)B_1(\lambda), \tag{62}$$

for almost every $\lambda \in \mathbb{C}$.

Proof. By (60), there exist real matrices $A_2 \in \mathbb{R}^{(n+m-k)\times n}$ and $B_2 \in \mathbb{R}^{(n+m-k)\times m}$ such that

$$\det \begin{bmatrix} A_1(\lambda) & B_1(\lambda) \\ A_2 & B_2 \end{bmatrix} \not\equiv 0. \tag{63}$$

Now let $T(\lambda) \in \mathbb{R}^{n \times k}(\lambda)$, $R(\lambda) \in \mathbb{R}^{n \times (n+m-k)}(\lambda)$ be defined by

$$\left[T(\lambda), R(\lambda)\right] = \left[A_1+\lambda A_{-1}, B_1+\lambda B_{-1}\right] \begin{bmatrix} A_1(\lambda) & B_1(\lambda) \\ A_2 & B_2 \end{bmatrix}^{-1}.$$

Then

$$T(\lambda)[A_1(\lambda), B_1(\lambda)] + R(\lambda)\,[A_2, B_2] = [A_1+\lambda A_{-1}, B_1+\lambda B_{-1}].$$

By (61), this implies

$$\ker\,[A_1(\lambda), B_1(\lambda)] \subset \ker R(\lambda)\,[A_2, B_2]$$

for almost every $\lambda \in \mathbb{C}$. Moreover, we have in any case

$$\ker\,[A_2, B_2] \subset \ker R(\lambda)\,[A_2, B_2].$$

Finally, it follows from (63) that

$$\ker\,[A_2, B_2] + \ker\,[A_1(\lambda), B_1(\lambda)] = \mathbb{C}^{n+m}$$

for almost every $\lambda \in \mathbb{C}$. Hence $R(\lambda)\,[A_2, B_2] \equiv 0$. □

Step 4 The solution $x(t)$ of Ω^T vanishes for $t \geq 0$.

Proof. It follows from (58) and (62) that

$$\left(\hat{x}^T(\lambda), \hat{\hat{x}}^T(\lambda)T(\lambda)\right)\begin{bmatrix} A_0-\lambda I & A_1+\lambda A_{-1} & B_0+\lambda B_{-0} & B_1+\lambda B_{-1} \\ A_1(\lambda) & 0 & B_1(\lambda) & 0 \end{bmatrix} \equiv x^T. \qquad (64)$$

Moreover, by (59) there exist unimodular matrices $M(\lambda)$, $N(\lambda)$ of appropriate size such that

$$M(\lambda) \begin{bmatrix} A_0-\lambda I & A_1+\lambda A_{-1} & B_0+\lambda B_{-0} & B_1+\lambda B_{-1} \\ A_1(\lambda) & 0 & B_1(\lambda) & 0 \end{bmatrix} N(\lambda)$$

$$= \begin{bmatrix} \alpha_1(\lambda) & & & 0 & \cdots & 0 \\ & \ddots & & \vdots & & \vdots \\ & & \alpha_{n+k}(\lambda) & 0 & \cdots & 0 \end{bmatrix}$$

is in Smith form where all the $\alpha_j(\lambda)$ are nonzero polynomials. Now let $\tilde{N}(\lambda)$ consist of the left $n + k$ columns of $N(\lambda)$. Then we obtain

$$\begin{bmatrix} A_0-\lambda I & A_1+\lambda A_{-1} & B_0+\lambda B_{-0} & B_1+\lambda B_{-1} \\ A_1(\lambda) & 0 & B_1(\lambda) & 0 \end{bmatrix} \tilde{N}(\lambda) \begin{bmatrix} \alpha_1(\lambda)^{-1} & & \\ & \ddots & \\ & & \alpha_{n+k}(\lambda)^{-1} \end{bmatrix} M(\lambda)$$

$$= I_{n+k}.$$

By (64), this implies

$$x^T \tilde{N}(\lambda) \begin{bmatrix} \alpha_1(\lambda)^{-1} & & \\ & \ddots & \\ & & \alpha_{n+k}(\lambda)^{-1} \end{bmatrix} M(\lambda) = \left(\hat{x}^T(\lambda), \hat{\hat{x}}^T(\lambda)T(\lambda)\right)$$

The function on the left-hand side is of exponential growth zero. Hence it follows from a theorem of Paley and Wiener (see, e.g., Rudin [131, Theorem 19.3]) that $\hat{x}(\lambda) \equiv 0$ and thus $x(t) = 0$ for $t \geq 0$. □

The criterion in the above theorem can be generalized to systems with commensurable delays, but we will not do this here. In a more general situation, the derivation of an analogous result seems to be a hard problem.

For retarded systems with undelayed input variables (i.e., $A_{-1} = 0$ and $B_{-0} = B_{-1} = B_1 = 0$), the criterion of Theorem 4.3.7 reduces to

$$\operatorname{rank} \begin{bmatrix} A_0-\lambda I & A_1 & B_0 \\ A_1 & 0 & 0 \end{bmatrix} = n + \operatorname{rank} A_1 \tag{65}$$

for some $\lambda \in \mathbb{C}$. This condition has been derived by Manitius [95]. Moreover it has been proved in [95] that (65) implies the existence of a feedback matrix $K \in \mathbb{R}^{m \times n}$ such that the closed loop system

$$\begin{aligned} \dot{x}(t) &= A_0x(t) + A_1x(t-h) + B_0u(t), \\ u(t) &= Kx(t), \end{aligned} \tag{66}$$

is F-complete. This is equivalent to

$$\operatorname{rank} \begin{bmatrix} A_0+B_0K-\lambda I & A_1 \\ A_1 & 0 \end{bmatrix} = n + \operatorname{rank} A_1 \tag{67}$$

for some $\lambda \in \mathbb{C}$ (Corollary 3.2.5). In the presence of input delays such a statement is meaningless since a feedback changes the structural operator F, even if there are no (additional) delays in the loop.

4.3.8 EXAMPLES

(i) We have seen that the scalar n-th order differential-difference equation (29) is spectrally controllable (Example 4.2.13 (i)) and F-complete (Example 3.2.6 (ii)). Hence (29) is F-controllable (Corollary 4.3.6).

(ii) Consider system (33) which is described by the matrices

$$A_0 = \begin{bmatrix} 0 & 0 \\ -1 & 0 \end{bmatrix}, \; A_1 = \begin{bmatrix} 1 & 0 \\ 0 & 0 \end{bmatrix}, \; A_{-1} = \begin{bmatrix} 0 & 1 \\ 0 & 0 \end{bmatrix},$$

$$B_0 = B_{-0} = B_{-1} = \begin{bmatrix} 0 \\ 0 \end{bmatrix}, \; B_1 = \begin{bmatrix} 1 \\ 0 \end{bmatrix},$$

(see Example 4.2.13 (iii)). This system is spectrally but not approximately controllable. However, condition (52) is satisfied since

$$\operatorname{rank}\begin{bmatrix} -\lambda & 0 & 1 & \lambda & 0 & 1 \\ -1 & -\lambda & 0 & 0 & 0 & 0 \\ 1 & \lambda & 0 & 0 & 1 & 0 \\ 0 & 0 & 0 & 0 & 0 & 0 \end{bmatrix} = 3$$

for every $\lambda \in \mathbb{C}$. Hence (33) is F-controllable.

Note that F-controllability of (33) will be destroyed, if the delay in the input disappears which means that the matrices B_0 and B_1 are interchanged.

(iii) The lossless transmission line

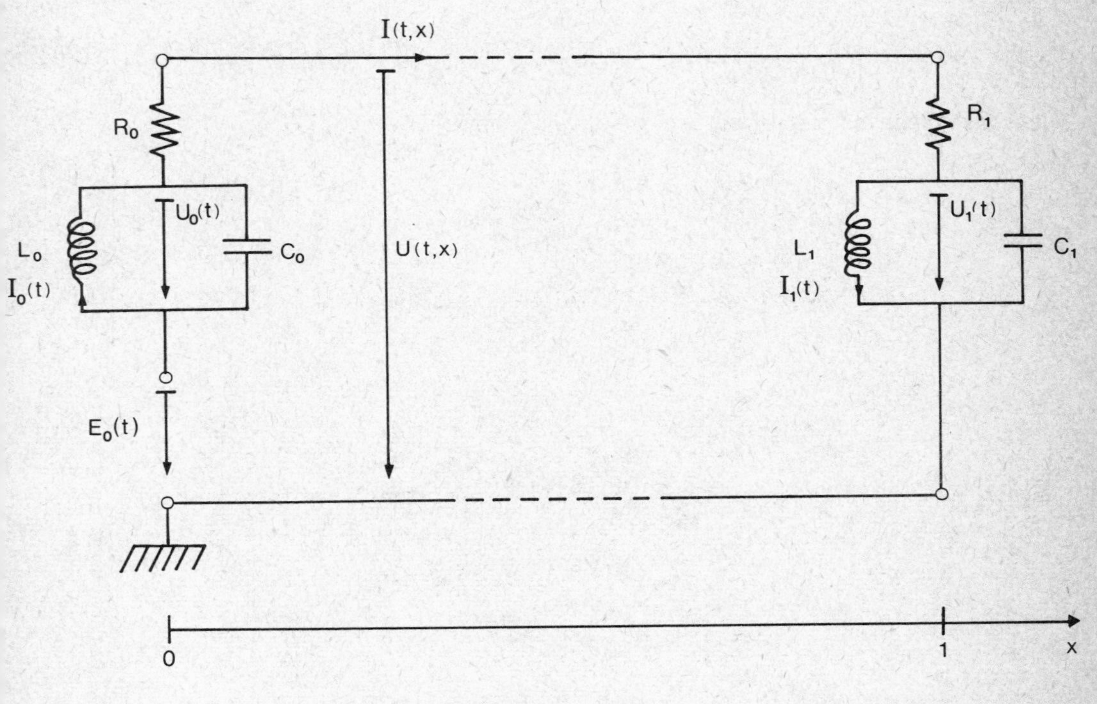

Figure 5

can be described by the hyperbolic PDE

$$\frac{\partial U}{\partial x} = - L \frac{\partial I}{\partial t}, \quad \frac{\partial I}{\partial x} = - C \frac{\partial U}{\partial t} \tag{68}$$

with boundary conditions

$$U(t,0) = U_0(t) - R_0 I(t,0) + E_0(t), \; U(t,1) = U_1(t) + R_1 I(t,1), \tag{69.1}$$

$$U_0(t) = -L_0 \dot{I}_0(t), \; U_1(t) = L_1 \dot{I}_1(t), \tag{69.2}$$

$$I(t,0) - I_0(t) = - C_0 \dot{U}_0(t), \; I(t,1) - I_1(t) = C_1 \dot{U}_1(t). \tag{69.3}$$

Integrating the PDE (68) along its characteristics we obtain

$$x_1(t) = \sqrt{C}\ U(t,0) + \sqrt{L}\ I(t,0) = \sqrt{C}\ U(t+h,1) + \sqrt{L}\ I(t+h,1),$$

$$x_2(t) = \sqrt{C}\ U(t,1) - \sqrt{L}\ I(t,1) = \sqrt{C}\ U(t+h,0) - \sqrt{L}\ I(t+h,0),$$

where $h = \sqrt{CL}$. Now let us introduce the additional variables $x_3(t) = 2\sqrt{L}\ I_0(t)$, $x_4(t) = 2\sqrt{L}\ I_1(t)$ and $u(t) = 2\sqrt{C}\ E_0(t)$. Then the boundary conditions (69) lead to a NFDE of the form

$$d/dt\ (x(t) - A_{-1}x(t-h) - B_{-0}u(t)) = A_0x(t) + A_1x(t-h) + B_0u(t). \quad (70)$$

The corresponding matrices are given by

$$A_0 = \begin{bmatrix} -\alpha_0 & 0 & \alpha_0 & 0 \\ 0 & -\alpha_1 & 0 & -\alpha_1 \\ -\alpha_2 & 0 & 0 & 0 \\ 0 & \alpha_3 & 0 & 0 \end{bmatrix},\ A_1 = \begin{bmatrix} 0 & \alpha_0 & 0 & 0 \\ \alpha_1 & 0 & 0 & 0 \\ 0 & \alpha_2\alpha_4 & 0 & 0 \\ -\alpha_3\alpha_5 & 0 & 0 & 0 \end{bmatrix},\ B_0 = \begin{bmatrix} 0 \\ 0 \\ \alpha_2\beta_0 \\ 0 \end{bmatrix}, \quad (71.1)$$

$$A_{-1} = \begin{bmatrix} 0 & \alpha_4 & 0 & 0 \\ \alpha_5 & 0 & 0 & 0 \\ 0 & 0 & 0 & 0 \\ 0 & 0 & 0 & 0 \end{bmatrix},\ B_{-0} = \begin{bmatrix} \beta_0 \\ 0 \\ 0 \\ 0 \end{bmatrix}, \quad (71.2)$$

where

$$\alpha_0 = \frac{1}{C_0}\frac{\sqrt{C}}{R_0\sqrt{C}+\sqrt{L}},\ \alpha_2 = \frac{1}{L_0}\frac{R_0\sqrt{C}+\sqrt{L}}{\sqrt{C}},\ \alpha_4 = \frac{R_0\sqrt{C}-\sqrt{L}}{R_0\sqrt{C}+\sqrt{L}},\ \beta_0 = \frac{\sqrt{L}}{R_0\sqrt{C}+\sqrt{L}}, \quad (71.3)$$

$$\alpha_1 = \frac{1}{C_1}\frac{\sqrt{C}}{R_1\sqrt{C}+\sqrt{L}},\ \alpha_3 = \frac{1}{L_1}\frac{R_1\sqrt{C}+\sqrt{L}}{\sqrt{C}},\ \alpha_5 = \frac{R_1\sqrt{C}-\sqrt{L}}{R_1\sqrt{C}+\sqrt{L}}. \quad (71.4)$$

It is easy to see that the corresponding free system satisfies the F-completeness criterion (3.13) as long as C, L, C_0, L_0, C_1, L_1 are nonzero and finite and R_0, R_1 are finite. Moreover, we have

$$\operatorname{rank}\left[\Delta(\lambda),\ B_0 + \lambda B_{-0}\right]$$

$$= \operatorname{rank}\begin{bmatrix} \lambda+\alpha_0 & -(\alpha_0+\lambda\alpha_4)e^{-\lambda h} & -\alpha_0 & 0 & \lambda\beta_0 \\ -(\alpha_1+\lambda\alpha_5)e^{-\lambda h} & \lambda+\alpha_1 & 0 & \alpha_1 & 0 \\ \alpha_2 & -\alpha_2\alpha_4 e^{-\lambda h} & \lambda & 0 & \alpha_2\beta_0 \\ \alpha_3\alpha_5 e^{-\lambda h} & -\alpha_3 & 0 & \lambda & 0 \end{bmatrix}$$

$$= \operatorname{rank}\begin{bmatrix} \alpha_0 e^{\lambda h} & -\alpha_0 e^{-\lambda h} & -\alpha_0 & 0 & \lambda\beta_0 \\ -\alpha_1\alpha_3 & \alpha_1\alpha_3 & 0 & \alpha_1\alpha_3+\lambda^2 & 0 \\ 0 & 0 & \lambda & 0 & \alpha_2\beta_0 \\ \alpha_3\alpha_5 & -\alpha_3 & 0 & \lambda & 0 \end{bmatrix}.$$

Hence spectral controllability fails in the resonance case

$$\alpha_0\alpha_2 = \alpha_1\alpha_3 = k^2\pi^2/h^2,\ k \in \mathbb{N},$$

which is equivalent to

$$C_0L_0 = C_1L_1 = CL/k^2\pi^2,\ k \in \mathbb{N}. \tag{72}$$

In this situation $\lambda = \pm\, i/\sqrt{C_0L_0}$ is an uncontrollable eigenvalue.

We conclude that system (70), (71) is F-controllable if and only if (72) does not hold (note that α_5 can never be equal to one). Moreover, if (72) is satisfied, then the system is not stabilizable.

If there is any distributed delay in the system, then we cannot apply Theorem 4.3.7. However, in some cases it is still possible to say something about F-controllability. We will do this in a final example.

4.3.10 EXAMPLE. We shall prove that the NFDE

$$d/dt\ (x_1(t) - x_3(t-2)) = x_1(t) + \int_{-1}^{0} x_2(t+\tau)d\tau \tag{73.1}$$

$$\dot{x}_2(t) = x_2(t-1) + \int_{-1}^{0} x_3(t+\tau)d\tau \tag{73.2}$$

$$\dot{x}_3(t) = \int_{-1}^{0} u(t+\tau)d\tau \tag{73.3}$$

is F-controllable, but not controllable in the sense of Definition 4.2.1.

(i) <u>Spectral controllability</u>. We have to show that the matrix

$$\begin{array}{cc} & \begin{array}{cccc} \quad 1 \quad & \quad 2 \quad & \quad 3 \quad & \quad 4 \quad \end{array} \\ [\Delta(\lambda)\ B(e^{\lambda\cdot})] = & \begin{bmatrix} \lambda-1 & \frac{e^{-\lambda}-1}{\lambda} & -\lambda e^{-2\lambda} & 0 \\ 0 & \lambda-e^{-\lambda} & \frac{e^{-\lambda}-1}{\lambda} & 0 \\ 0 & 0 & \lambda & \frac{1-e^{-\lambda}}{\lambda} \end{bmatrix} \end{array}$$

is of rank 3 for every $\lambda \in \mathbb{C}$.

<u>$\lambda = 0$</u> Note that $\lim\limits_{\lambda\to 0} \lambda^{-1}(1-e^{-\lambda}) = 1$. Hence Columns 1, 2 and 4 are linearly independent.

<u>$\lambda = 1$</u> Columns 2, 3 and 4 are linearly independent.

<u>$\lambda \in i\mathbb{R},\ \lambda \neq 0$</u> In this case $\lambda \neq e^{-\lambda}$ and hence Columns 1, 2 and 3 are linearly independent.

<u>$\lambda \notin i\mathbb{R},\ \lambda \neq 1$</u> In this case $e^{-\lambda} \neq 1$ and hence Columns 1, 3 and 4 are linearly independent.

(ii) <u>F-controllability</u>. We have to show that the nontrivial small solutions of the transposed system

$$\dot{x}_1(t) = x_1(t), \tag{74.1}$$

$$\dot{x}_2(t) = \int_{-1}^{0} x_1(t+\tau)d\tau + x_2(t-1), \tag{74.2}$$

$$\dot{x}_3(t) = \dot{x}_1(t-2) + \int_{-1}^{0} x_2(t+\tau)d\tau, \tag{74.3}$$

$$y(t) = \int_{-1}^{0} x_3(t+\tau)d\tau, \tag{74.4}$$

are observable (Theorem 4.3.5). For this purpose let $x(t)$, $t \geq -2$, be a solution of (74) which vanishes for $t \geq 1$, and suppose that the corresponding output $y(t)$, $t \geq 0$, is identically zero.

Then $0 = \dot{y}(t) = x_3(t) - x_3(t-1)$ for $t \geq 0$ and hence

$$x_3(t) = 0 \ \forall \ t \geq -1.$$

By (74.3), this implies

$$\dot{x}_1(t) = - \int_{t+1}^{t+2} x_2(s)ds, \ t \geq -2,$$

and thus

$$x_1(t) = 0 \ \forall \ t \geq 0.$$

Finally, it follows from (74.2) that

$$x_2(t) = \dot{x}_2(t+1) - \int_{t}^{t+1} x_1(s)ds = 0, \ \forall \ t \geq 0.$$

(iii) Approximate controllability fails. Let $x_1(t) = x_3(t) = 0$ for $t \geq -2$, and let $x_2(t)$ be nonzero for $-2 \leq t < -1$ and zero for $t \geq -1$. Then it is easy to see that $x(t)$ is a small solution of (74) with zero output for $t \geq -1 = h_u - h_x$. Hence the small solutions of (74) are not observable. We conclude that (74) is not strictly observable (Proposition 4.2.7) and hence (73) is not approximately controllable (Theorem 4.2.6).

CONCLUSIONS

At the end of this chapter let us briefly review the controllability properties of the NFDE

$$d/dt\ (x(t) - A_{-1}x(t-h)) = A_0x(t) + A_1x(t-h) + B_0u(t). \tag{75}$$

In Section 4.2 we have shown that this system is approximately controllable in the state space M^p (or equivalently in $W^{1,p}$) if and only if

$$\text{rank} \left[\Delta(\lambda), B_0\right] = n \quad \forall \ \lambda \in \mathbb{C}, \tag{a}$$

$$\max_{\lambda \in \mathbb{C}} \text{rank} \left[A_1 + \lambda A_{-1}, B_0\right] = n, \tag{b}$$

(see also Bartosiewicz [10], O'Connor-Tarn [110]).

Motivated by several examples and by the work of Manitius [91, 95] on retarded systems, we have introduced in Section 4.3 the weaker concept of (approximate) F-controllability. We have shown that, for the system (75), this concept is equivalent to condition (a) and

$$\max_{\lambda\in\mathbb{C}} \operatorname{rank} \begin{bmatrix} A_0-\lambda I & A_1+\lambda A_{-1} & B_0 \\ A_1+\lambda A_{-1} & 0 & 0 \end{bmatrix} = n + \max_{\lambda\in\mathbb{C}} \operatorname{rank} [A_1+\lambda A_{-1}]. \tag{b'}$$

On the other hand, Example 4.1.12 shows that a system of the form (75) may be approximately controllable even though it is not stabilizable through a feedback of the form

$$u(t) = K_{-1}\dot{x}(t-h) + K_0x(t) + K_1x(t-h) + \int_{-h}^{0} K_{01}(\tau)x(t+\tau)d\tau \tag{76}$$

(O'Connor-Tarn [111], see also Chapter 5). Does this mean that weakening the concept of approximate controllability was a step in the wrong direction? We guess that it was not! To be more precise, we note that system (75) is stabilizable through a feedback of the form (76) (with arbitrary decay rate) if and only if condition (a) is satisfied and

$$\operatorname{rank} [\lambda I-A_{-1}, B_0] = n \quad \forall\ \lambda \in \mathbb{C},\ \lambda \neq 0, \tag{c}$$

(Section 5.1, O'Connor-Tarn [111]). Condition (c) means that the *nonzero* eigenvalues of A_{-1} are controllable via the input matrix B_0. On the other hand, controllability of the *zero* eigenvalue of A_{-1} is equivalent to

$$\operatorname{rank} [A_{-1}\ B_0] = n. \tag{d}$$

This condition is stronger than (b). Hence the conditions (c) and (b) are independent. In other words, controllability of the eigenvalue $\lambda = 0$ of A_{-1} is completely unimportant for the purpose of stabilization. The effect of feedback may be even to make all the eigenvalues of $A_{-1} + B_0K_{-1}$ equal to zero by choice of K_{-1} (dead beat control). Hence the weakening of condition (b) does not affect the feedback stabilization properties of system (75).

The question remains as to whether or not condition (c) has something to do with the state space properties of system (75). To discover this, let us take a look at the stronger condition

$$\text{rank } [\lambda I - A_{-1} \ B_0] = n \ \forall \lambda \in \mathbb{C} \tag{e}$$

which is equivalent to (c) and (d). It has been proved by Jakubczyk [62] that (e) is satisfied if and only if the reachable subspace R of (75) is closed in $W^{1,p}$ and has a finite codimension. This shows that (a) and (e) together are equivalent to exact controllability of system (75) in the state space $W^{1,p}$ (see also Rodas-Langenhop [130], Bartosiewicz [10, Proposition 16], O'Connor-Tarn [110, Corollary 5.8]).

CONJECTURE System (75) is exactly null-controllable in the state space $W^{1,p}$ if and only if (a) and (c) are satisfied.

5 Feedback stabilization and dynamic observation

The problems of feedback stabilization and dynamic observation for retarded systems with undelayed input/output variables have been widely studied by various authors. For the feedback problem we refer to Krasovskii [76], Osipov [123], Krasovskii-Osipov [78], Pandolfi [125], Manitius [91], Olbrot [118] and Olbrot-Gasiewski [120]. The dual problem of designing a Luenberger-type observer for retarded systems with undelayed output variables has been investigated, for instance, by Gressang [40], Gressang-Lamont [41], Hewer-Nazaroff [51], Bhat-Koivo [12], Olbrot [119], Salamon [132] and Lee-Olbrot [83]. Some duality results between these two concepts can be found in Salamon [135].

The feedback problem for systems with control delays only has been treated by Olbrot [113], Manitius-Olbrot [97], and - by different methods - by Kwon-Pearson [80]. Watanabe-Ito [148] and Klamka [72] have constructed a dynamic compensator for systems with delays in control and observation. A stabilizing control law and a dynamic observer for retarded systems with general delays in input, state and output has been developed in Salamon [133].

Some research effort has also been devoted to solving the feedback stabilization problem and designing a dynamic observer for delay systems within the algebraic theory of systems over rings. Results in this direction can be found, for instance, in Kamen [66], Morse [108], Sontag [141], Hazewinkel [45] and Hautus-Sontag [44].

Only very little work has been done on feedback stabilization of neutral systems. The only papers in this area seem to be those of Pandolfi [126], Jakubczyk-Olbrot [63] and O'Connor-Tarn [111]. These authors do not allow delays in the input variables. Moreover, the assumptions in Jakubczyk-Olbrot [63] are rather storng, and the main result in Pandolfi [126] (infinite pole-shifting) is wrong. Some interesting ideas can be found in O'Connor-Tarn [111].

Apparently, there are no results on dynamic observation of NFDEs in the open literature on delay systems.

5.1 PRELIMINARIES

The main problem in stabilizing a NFDE - in comparison with the retarded case - is the fact that there may exist an infinite number of unstable eigenvalues. Therefore a neutral system has to be stabilized in two steps. First, one has to apply a control law which guarantees that there are only finitely many unstable eigenvalues left. This means stabilization of the difference equation (see O'Connor-Tarn [111] for systems with a single point delay). In a second step, the resulting closed-loop system can be stabilized by finite pole-shifting (see Pandolfi [126] for NFDEs with state delays only).

Before going into detail, let us discuss first the problem of stability. It is in general not known if the asymptotic behaviour of the semigroups $S(t)$ and $\mathcal{S}(t)$ (introduced in Section 2.1) is completely determined by the spectrum of the generator. Therefore we have to restrict ourselves to the case that the function $\mu: [-h,0] \to \mathbb{R}^{n\times n}$ of bounded variation contains no singular part. This means that $\mu(\tau)$ can be written in the form

$$\mu(\tau) = - \sum_{j=1}^{\infty} A_{-j}\chi_{(-\infty,-h_j]}(\tau) - \int_{\tau}^{0} A_{-\infty}(\sigma)d\sigma, \ -h \leqslant \tau \leqslant 0, \tag{1}$$

where $0 < h_j \leqslant h$ for $j \in \mathbb{N}$, $\chi_{(-\infty,-h_j]}$ denotes the indicator function of the interval $(-\infty,-h_j]$, and the matrices A_{-j} satisfy

$$\sum_{j=1}^{\infty} ||A_{-j}|| + \int_{-h}^{0} ||A_{-\infty}(\tau)|| \, d\tau < \infty . \tag{2}$$

The bounded, linear functional $M : C \to \mathbb{R}^n$ is then given by

$$M\phi = \sum_{j=1}^{\infty} A_{-j}\phi(-h_j) + \int_{-h}^{0} A_{-\infty}(\tau)\phi(\tau)d\tau, \ \phi \in C. \tag{3}$$

In this case it has been proved by Henry [50] that the exponential growth of the semigroup $\mathcal{S}_C(t) : C \to C$ is in fact determined by the spectrum of its generator (see also Hale [42, Section 12.10]). The same arguments apply to the semigroups $S(t)$ and $\mathcal{S}(t)$ (compare Theorem 5.2.7 below). More precisely, we have the following result.

5.1.1 <u>THEOREM</u> (Henry)

Let $M : C \to \mathbb{R}^n$ *be given by* (3), (2). *Then*

$$\omega_0 = \lim_{t\to\infty} t^{-1} \log \|S(t)\|_{L(M^p)}$$

$$= \lim_{t\to\infty} t^{-1} \log \|S(t)\|_{L(W^{1,p})}$$

$$= \sup \{\mathrm{Re}\ \lambda | \lambda \in \sigma(A)\}.$$

There is another reason for restricting the discussion of this chapter to the case that M is given by (3). This is the fact that the sequences of eigenvalues of A with bounded real part are already determined by the difference equation

$$\Sigma_0 \qquad x(t) = \sum_{j=1}^{\infty} A_{-j} x(t-h_j).$$

The eigenvalues of Σ_0 are characterized by the complex matrix function

$$\Delta_0(\lambda) = I - \sum_{j=1}^{\infty} A_{-j} e^{-\lambda h_j}. \tag{4}$$

The relation between the eigenvalues of Σ and those of Σ_0 is explained in the following lemma which can be obtained by combining Hale-Meyer [43, p. 37, Lemma 1] with Henry [50, Lemma 3.2]. For completeness, we give a proof of this result.

5.1.2 <u>LEMMA</u>. *Let $\alpha < \beta$ be given. Then the following statements are equivalent.*

(i) *There exists some $\lambda_0 \in \mathbb{C}$ such that $\alpha < \mathrm{Re}\ \lambda_0 < \beta$ and $\det \Delta_0(\lambda_0) = 0$.*

(ii) *There exists some $\varepsilon > 0$ and a sequence $\lambda_k \in \mathbb{C}$ such that $|\mathrm{Im}\ \lambda_k|$ tends to infinity, $\alpha + \varepsilon \leqslant \mathrm{Re}\ \lambda_k \leqslant \beta-\varepsilon$, and $\det \Delta(\lambda) = 0$ for every $k \in \mathbb{N}$.*

<u>Proof</u>. First note that

$$\lambda^{-1}\Delta(\lambda) = \Delta_0(\lambda) - \int_{-h}^{0} A_{-\infty}(\tau)e^{\lambda\tau}d\tau - \lambda^{-1}L(e^{\lambda\cdot}),\ \lambda \in \mathbb{C},$$

and hence the limit

$$\lim_{|\mathrm{Im}\lambda| \to \infty} |\lambda^{-n} \det \Delta(\lambda) - \det \Delta_0(\lambda)| = 0 \tag{5}$$

exists uniformly for $\alpha \leqslant \mathrm{Re}\,\lambda \leqslant \beta$ (this follows from the lemma of Riemann-Lebesgue). Moreover, $\det \Delta_0(\lambda)$ is an almost periodic function in the strip $\alpha \leqslant \mathrm{Re}\,\lambda \leqslant \beta$ (see, e.g., Bohr [15], Corduneanu [23]).

'(i) ⇒ (ii)' Choose $\varepsilon > 0$ such that $\alpha + 2\varepsilon < \mathrm{Re}\,\lambda_0 < \beta - 2\varepsilon$ and the implication

$$0 < |\lambda - \lambda_0| \leqslant \varepsilon \Rightarrow \det \Delta_0(\lambda) \neq 0$$

holds for every $\lambda \in \mathbb{C}$. Also define

$$\delta = \inf_{|\lambda-\lambda_0| = \varepsilon} |\det \Delta_0(\lambda)| > 0.$$

Then, by (5), there exists some constant $c > 0$ such that the inequality

$$|\lambda^{-n} \det \Delta(\lambda) - \det \Delta_0(\lambda)| \leqslant \delta/3 \tag{6}$$

holds for every $\lambda \in \mathbb{C}$ which satisfies $\alpha \leqslant \mathrm{Re}\,\lambda \leqslant \beta$ and $|\mathrm{Im}\,\lambda| \geqslant c$. Since $\det \Delta_0(\lambda)$ is almost periodic in the strip $\alpha \leqslant \mathrm{Re}\,\lambda \leqslant \beta$, there exists also a sequence c_k of real numbers tending to infinity and satisfying

$$\begin{aligned} &|c_k + \mathrm{Im}\,\lambda_0| \geqslant c + \varepsilon, \\ &|\det \Delta_0(\lambda) - \det \Delta_0(\lambda - ic_k)| \leqslant \delta/3, \end{aligned} \tag{7}$$

for every $k \in \mathbb{N}$ and every $\lambda \in \mathbb{C}$, $\alpha \leqslant \mathrm{Re}\,\lambda \leqslant \beta$.

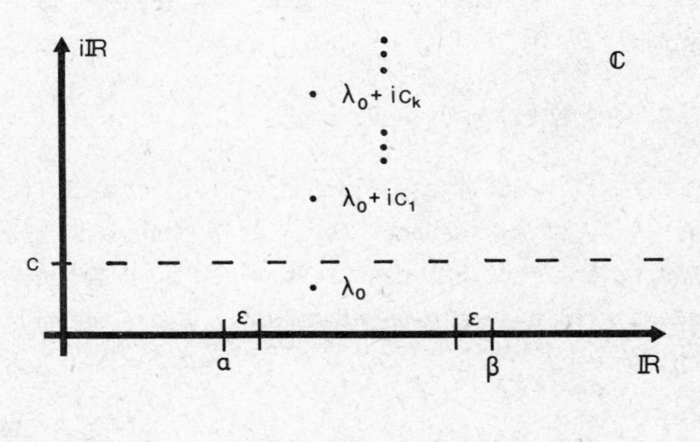

Figure 6

By (6) and (7), the following inequality holds for every $k \in \mathbb{N}$ and every $\lambda \in \mathbb{C}$ which satisfies $|\lambda-\lambda_0-ic_k| = \varepsilon$.

$$|\lambda^{-n} \det \Delta(\lambda) - \det \Delta_0(\lambda-ic_k)|$$

$$\leqslant 2\delta/3 < \delta = \inf_{|\lambda-\lambda_0-ic_k| = \varepsilon} |\det \Delta_0(\lambda-ic_k)|.$$

Hence it follows from Rouché's theorem that, for every $k \in \mathbb{N}$, there exists some $\lambda_k \in \mathbb{C}$ such that $|\lambda_k-\lambda_0-ic_k| < \varepsilon$ and $\det \Delta(\lambda_k) = 0$.

'(ii) $\Rightarrow$ (i)' Suppose that $\det \Delta_0(\lambda) \neq 0$ for every $\lambda \in \mathbb{C}$ which satisfies $\alpha < \text{Re}\,\lambda < \beta$. Moreover let $\varepsilon > 0$ and define

$$\delta = \inf_{\alpha+\varepsilon \leqslant \text{Re}\lambda \leqslant \beta-\varepsilon} |\det \Delta_0(\lambda)|.$$

Then it follows from a result of Lewin [86, p. 267] that $\delta > 0$. Applying again equation (5), we conclude that $\det \Delta(\lambda) \neq 0$ for every $\lambda \in \mathbb{C}$ with $\alpha + \varepsilon \leqslant \text{Re}\,\lambda \leqslant \beta-\varepsilon$ and sufficiently large imaginary part. This contradicts (ii). □

The above result has important consequences for the feedback stabilization of system Ω where M is given by (3) and B,Γ by

$$B\xi = B_0\xi(0), \; \Gamma\xi = 0, \; \xi \in C([-h,0];\mathbb{R}^m), \tag{8}$$

(no input delays). By Lemma 5.1.2, the uncontrolled system has infinitely many unstable eigenvalues - including sequences with real part tending to zero - if and only if

$$\sup \{\text{Re}\,\lambda | \lambda \in \mathbb{C}, \det \Delta_0(\lambda) = 0\} \geqslant 0.$$

Hence a simultaneous shifting of these eigenvalues to a stable region of the form $\{\lambda \in \mathbb{C} | \text{Re}\,\lambda \leqslant -\varepsilon\}$ by state feedback requires a change of the difference equation Σ_0 (this important fact has been recognized by O'Connor-Tarn [111, Theorem 2] for systems with a single point delay). Hence we have to allow control laws of the form

$$u(t) = \sum_{j=1}^{\infty} K_{-j}\dot{x}(t-h_j) + \int_{-h}^{0} K_{-\infty}(\tau)\dot{x}(t+\tau)d\tau + Kx_t \tag{9}$$

for system Ω. We assume that K is a bounded linear functional on C with values in $\mathbb{R}^m$ and that the matrices K_{-j} satisfy

$$\sum_{j=1}^{\infty} \|K_{-j}\|_{\mathbb{R}^{m\times n}} + \int_{-h}^{0} \|K_{-\infty}(\tau)\|_{\mathbb{R}^{m\times n}} d\tau < \infty . \tag{10}$$

These observations show that the infinite pole-shifting result of Pandolfi [126] is wrong, since the control law in [126] does not change the difference part of the equation.

By Theorem 5.1.1, the closed-loop system Ω,(9) is exponentially stable if and only if there exists some $\varepsilon > 0$ such that

$$\det \left\{\Delta(\lambda) - B_0 \left[\sum_{j=1}^{\infty} \lambda K_{-j} e^{-\lambda h_j} + \int_{-h}^{0} \lambda e^{\lambda\tau} K_{-\infty}(\tau) d\tau + K(e^{\lambda\cdot})\right]\right\} \neq 0$$

for every $\lambda \in \mathbb{C}$ with Re $\lambda \geqslant -\varepsilon$. By Lemma 5.1.2, this implies that

$$\det \left[\Delta_0(\lambda) - B_0 \sum_{j=1}^{\infty} K_{-j} e^{-\lambda h_j}\right] \neq 0$$

for every $\lambda \in \mathbb{C}$ with Re $\lambda \geqslant -\varepsilon$. Hence, we obtain two necessary conditions for stabilizability.

5.1.3 <u>COROLLARY</u>. *Let* M *be given by* (3) *and* B, Γ *by* (8). *Moreover, suppose that system* Ω *can be made exponentially stable through a control law of the form* (9). *Then there exists some* $\varepsilon > 0$ *such that*

$$\text{rank } [\Delta(\lambda), B_0] = n \ \forall \ \lambda \in \mathbb{C}, \text{ Re } \lambda \geqslant -\varepsilon, \tag{11}$$

$$\text{rank } [\Delta_0(\lambda), B_0] = n \ \forall \ \lambda \in \mathbb{C}, \text{ Re } \lambda \geqslant -\varepsilon, \tag{12}$$

The necessity of condition (11) has already been proved by Pandolfi [126] and the necessity of (12) by O'Connor-Tarn [111, Theorem 3.1] for systems with a single point delay. The following examples show that these two conditions are independent.

5.1.4 <u>EXAMPLES</u>. Consider the NFDE Ω where L and M are given by

$$L\phi = A_0\phi(0) + A_1\phi(-h), \tag{13.1}$$

$$M\phi = A_{-1}\phi(-h), \ \phi \in C, \tag{13.2}$$

and B, Γ by (8). Then condition (12) is equivalent to rank $[I-A_{-1}e^{-\lambda h}, B_0] = n$ for all $\lambda \in \mathbb{C}$ with Re $\lambda \geqslant 0$. This means that

$$\operatorname{rank}\,[sI-A_{-1}, B_0] = n \;\forall\; s \in \mathbb{C},\ |s| \geqslant 1. \tag{14}$$

(i) If the matrices A_0, A_1, A_{-1}, B_0 are as in example 4.1.12, then system Ω is spectrally (even approximately) controllable, but (14) is not satisfied.

(ii) If $A_0 = A_1 = 0$, $m < n$, and if the matrix pair (A_{-1}, B_0) is controllable, then (14) is satisfied, but the eigenvalue $\lambda = 0$ of system Ω is not controllable.

The question remains as to whether or not (11) and (12) are also sufficient for the stabilizability of system Ω. One may expect that condition (12) guarantees the existence of a stabilizing control law for the difference equation Σ_0 with the additional input $B_0u(t)$.

STABILIZATION OF THE DIFFERENCE EQUATION

Let us consider the case that μ has only finitely many jumps such that the difference equation Σ_0 is of the form

$$\Sigma_0 \qquad x(t) = \sum_{j=1}^{N} A_{-j}x(t-h_j) + B_0u(t).$$

Moreover, we will first focus our attention on the extreme situation that no two delays are rationally independent. Then we can assume that $h_j = \alpha j$, $j = 1,\dots,N$, for some $\alpha > 0$ (commensurable delays). In this case (12) is equivalent to

$$\operatorname{rank}\left[s^nI - \sum_{j=1}^{N} s^{n-j}A_{-j}\ \ B_0\right] = n \;\forall\; s \in \mathbb{C},\ |s| \geqslant 1, \tag{15}$$

and hence to

$$\operatorname{rank}\begin{bmatrix} sI-A_{-1} & -A_{-2} & & & -A_{-N} & B_0 \\ -I & sI & & & & 0 \\ & & \cdot & & & \cdot \\ & & & \cdot & & \cdot \\ & & & \cdot & & \cdot \\ & & & \cdot & & \cdot \\ & & & -I & sI & 0 \end{bmatrix} = nN$$

for every $s \in \mathbb{C}$, $|s| \geqslant 1$. This condition is satisfied if and only if there exist feedback matrices $K_{-1},\ldots,K_{-N} \in \mathbb{R}^{m\times n}$, such that the eigenvalues of the block matrix

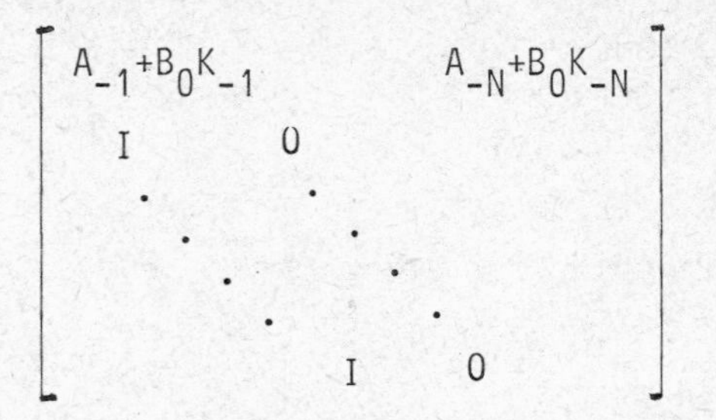

are inside the unit circle. It is easy to see that this is equivalent to the stability of the closed-loop system Σ_0 with the control law

$$u(t) = \sum_{j=1}^{N} K_{-j}\, x(t-h_j). \tag{16}$$

We conclude that, in the case of commensurable delays, condition (12) is in fact equivalent to the stabilizability of the difference equation Σ_0.

The derivation of an analogous result for systems with rationally independent delays seems to be a difficult open problem. But note that a statement in this form is only useful if all delays are fixed and known exactly, and if the delays can be determined precisely in the loop (16). These assumptions will be rather unrealistic, in general. In the applications one can assume that not all the independent parameters are known exactly. However, not all the parameters will be free in any case. For example, a 'shunted transmission line' (Hale [42, Section 12.5]) may lead to a scalar equation of the form Σ_0 where $N = 3$ and $h_3 = h_1 + h_2$. In this situation one should allow variations of the delays h_1 and h_2, but the third delay is always the sum of these two.

If there are any two independent delays, then we have to deal with the difficulty that the stability of Σ_0 is highly sensitive with respect to variations in the delays (see e.g., Melvin [103], Henry [50], Hale [42, Section 12.5], Carvalho [20]). Contrary to this sensitivity, stability is not affected by small variations in the coefficient matrices A_{-j}.

In the extreme case that all delays are independent, it should be the goal of feedback to make the system Σ_0 *strongly stable*, i.e., stable for any choice of the delays $h_1,\ldots,h_N$. A characterization of strong stability has been given by Hale [42, Section 12.5, Theorem 5.1]. Hale has shown that system Σ_0 is strongly stable if and only if it is stable for some fixed, rationally independent set of delays $\{h_1,\ldots,h_N\}$, and that this is equivalent to

$$r\left(\Sigma_0\right) := \sup \left\{ r\left(\sum_{j=1}^{N} A_{-j} e^{i\theta_j} \right) \;\middle|\; 0 \leqslant \theta_j \leqslant 2\pi \right\} < 1 \tag{17}$$

where $r(T)$ denotes the spectral radius of a matrix T. A necessary condition for strong stabilizability is that (12) holds for every set of delays $h_1,\ldots,h_N$ (ε depending on the h_j). It seems to be a difficult open question whether this condition is also sufficient.

Again, there is the additional difficulty that the delays in the feedback loop must be determined precisely in order to compensate any of the matrices A_{-j}. If this is not possible, i.e., if the delays in the feedback loop are allowed to vary (within a certain tolerance) independently of the delays in the given equation, then every feedback term B_0K_{-j} has to be treated as an additional term in the difference equation Σ_0. Now the following lemma shows that $r(\Sigma_0)$, as defined by (17), becomes larger with every additional term. We conclude, in the case of independent delays in the given equation and in the feedback loop, any control law of the form (16) leads to a worse stability behaviour of the difference equation Σ_0.

The proof of the following result has been communicated to the author personally by U. Helmke.

5.1.5 <u>LEMMA</u>. *Let* A, T $\in \mathbb{C}^{n\times n}$ *be given. Then* $r(A) \leqslant \sup_{|s|=1} r(A+sT)$.

<u>Proof</u>. Suppose that

$$\sup_{|s|=1} r(A+sT) < r(A).$$

Moreover, note that

$$r(A+sT) = \lim_{k\to\infty} \|(A+sT)^k\|^{1/k}.$$

Hence it follows from the compactness of the unit circle in the complex plane that there exists some $k \in \mathbb{N}$ such that

$$\|(A+sT)^k\| < \|A^k\| \quad \forall\, s \in \mathbb{C},\ |s| = 1.$$

This is a contradiction of the maximum principle (Hille-Phillips [53, Theorem 3.13.1]) applied to the holomorphic matrix function $s \to (A+sT)^k$. □

Let us now suppose that the difference equation Σ_0 is exponentially stable. Then system Σ has only finitely many unstable eigenvalues and we can apply the finite pole-shifting method of Pandolfi [126] for NFDEs with general delays in the state variable and undelayed input variables. In Section 5.3 we prove an analogous statement for systems with general delays in state and input as well as the dual result for NFDEs with output delays. To do this we have to study the perturbed semigroups which are obtained by applying the results of Section 1.3 to the systems Ω^T, Σ^T, $\tilde{\Sigma}$, $\tilde{\Omega}$. This will be done in the next section.

5.2 THE PERTURBED SEMIGROUPS

Throughout the rest of this chapter we will always assume that $\Gamma = 0$. Moreover, for the stability results, we need the assumption that M is given by (3).

This section is devoted to the study of the (perturbed) semigroups arising in state feedback and dynamic observation of a NFDE. On the level of the state space description we will discuss the closed loop feedback system only within the dual state concept (forcing terms). Hence we will work with the systems $\tilde{\Omega}$ and $\tilde{\Sigma}$ in the state spaces M^p and $W^{-1,p}$. One reason for this choice is that we have established the infinite-dimensional variation-of-constants formulae in Theorem 2.3.6. Correspondingly, the observer semigroup will be introduced within the original state concept (initial functions) represented by the systems Σ^T and Ω^T in the state spaces M^q and $W^{1,q}$. Let us begin with the observer semigroup of system Σ^T.

THE OBSERVER SEMIGROUP

Recall that the state $(z(t),x_t) \in M^q$ of system Σ^T is described by the semigroup $S^T(t)$. The corresponding output operator $\mathcal{B}^T : W^{1,q} \to \mathbf{R}^m$ is given by

$$\mathcal{B}^T\psi = \int_{-h}^{0} d\beta^T(\tau)\psi(\tau), \ \psi \in W^{1,q},$$

and cannot be extended to an operator on M^q, in general. However, this output operator satisfies the hypothesis (H3) of Section 1.3 which means that for every $T > 0$ there exists some $b_T > 0$ such that

$$\|\mathcal{B}^T S^T(\cdot)\psi\|_{L^q([0,T];\mathbb{R}^m)} \leq b_T \ \|\iota^T\psi\|_{M^q}$$

for every $\psi \in W^{1,q}$ (Remark 2.3.4 (ii)). This allows us to apply the theory of Section 1.3 in order to obtain an observer semigroup $S_K^T(t)$ in the state space M^q. Therefore we permit output injection operators $K^T : \mathbb{R}^m \to M^q$ given by

$$K^T y = (K_0^T y, K_1^T(\cdot)y) \in M^q, \ y \in \mathbb{R}^m,$$

where $K_0 \in \mathbb{R}^{m\times n}$ and $K_1(\cdot) \in L^q([-h,0];\mathbb{R}^{m\times n})$.

By Theorem 1.3.9, there exists a unique C_0-semigroup $S_K^T(t) : M^q \to M^q$ such that the following equation holds for every $\psi \in W^{1,q}$ and every $t > 0$

$$S_K^T(t)\iota^T\psi = S^T(t)\iota^T\psi + \int_0^t S_K^T(t-s)K^T\mathcal{B}^T S^T(s)\psi ds. \tag{18}$$

The infinitesimal generator of this semigroup is given by

$$\operatorname{dom} A_K^T = \operatorname{ran} \iota^T,$$

$$A_K^T\iota^T\psi = A^T\iota^T\psi + K^T\mathcal{B}^T\psi, \ \psi \in W^{1,q},$$

or explicitly

$$\begin{aligned} \operatorname{dom} A_K^T &= \{\psi \in M^q | \psi^1 \in W^{1,q}, \ \psi^0 = \psi^1(0) - M^T\psi^1\}, \\ A_K^T\psi &= (L^T\psi^1 + K_0^T\mathcal{B}^T\psi^1, \ \dot{\psi}^1 + K_1^T(\cdot)\mathcal{B}^T\psi^1), \end{aligned} \tag{19}$$

(Theorem 1.3.9 (ii), (iii)). This operator is not of the same type as A^T unless $K_1(\cdot) \equiv 0$ or $\mathcal{B}^T = 0$. Hence the semigroup $S_K^T(t)$ does not correspond to any neutral system of the form Σ^T, in general (compare Salamon [132] for retarded systems in the state space C). However, we will show that $S_K^T(t)$ can be regarded as a state space description for a certain system of the type (1.13).

We make use of the abbreviation

$$K_1^T * \zeta(\tau) = \int_\tau^0 K_1^T(\tau-\sigma)\zeta(\sigma)d\sigma, \ -h \leq \tau \leq 0.$$

5.2.1 THEOREM. *Let* $y(\cdot) \in L^q_{loc}([0,\infty);\mathbb{R}^m)$ *be given and let the triple* $z \in W^{1,q}_{loc}([0,\infty);\mathbb{R}^n)$, $x \in L^q_{loc}([-h,\infty);\mathbb{R}^n)$, $v \in L^q_{loc}([-h,\infty);\mathbb{R}^m)$ *satisfy the equations*

$$\Sigma^T_K \qquad \boxed{\begin{aligned} \dot{z}(t) &= L^T(x_t + K_1^T * v_t) + K_0^T v(t) \\ x(t) &= z(t) + M^T(x_t + K_1^T * v_t) \\ v(t) &= B^T(x_t + K_1^T * v_t) - y(t) \end{aligned}}$$

for $t \geq 0$. *Then*

$$\hat{x}(t) = (z(t), x_t + K_1^T * v_t) \in M^q \tag{20}$$

is given by the variation-of-constants formula

$$\hat{x}(t) = S_K^T(t)\hat{x}(0) - \int_0^t S_K^T(t-s)K^T y(s)ds. \tag{21}$$

Proof. Let

$$\tilde{S}_K^T(t) : M^q \times L^q \to M^q \times L^q$$

denote the C_0-semigroup which describes the evolution of the state $(z(t),x_t,v_t) \in M^q \times L^q$ of the homogeneous system Σ_K^T, i.e., $y(t) \equiv 0$ (Corollary 1.2.4). Then the infinitesimal generator of $\tilde{S}_K^T(t)$ is given by

$$\text{dom}\, \tilde{A}_K^T = \{(\psi,\zeta) \in M^q \times L^q | \psi^1 \in W^{1,q},\ \zeta \in W^{1,q}([-h,0];\mathbb{R}^m),$$

$$\psi^0 = \psi^1(0) - M^T(\psi^1 + K_1^T * \zeta),\ \zeta(0) = B^T(\psi^1 + K_1^T * \zeta)\},$$

$$\tilde{A}_K^T(\psi,\zeta) = (L^T(\psi^1 + K_1^T * \zeta) + K_0^T\zeta(0),\ \dot{\psi}^1,\ \dot{\zeta}),$$

(Theorem 1.2.6). Moreover, we introduce the bounded linear operator $T : M^q \times L^q \to M^q$ by defining

$$T(\psi,\zeta) = (\psi^0, \psi^1 + K_1^T * \zeta),\ \psi \in M^q,\ \zeta \in L^q.$$

Now let $(\psi,\zeta) \in \text{dom}\, \tilde{A}_K^T$ and $\Psi = T(\psi,\zeta) \in M^q$. Then $\Psi^1 \in W^{1,q}$ and

$$\Psi^0 = \psi^0 = \psi^1(0) - M^T(\psi^1 + K_1^T * \zeta) = \Psi^1(0) - M^T\Psi^1.$$

Hence $\Psi \in \operatorname{ran} \iota^T = \operatorname{dom} A_K^T$. Moreover, by (19),

$$[A_K^T\Psi]^0 = L^T\Psi^1 + K_0^T\mathcal{B}^T\Psi^1 = L^T(\psi^1 + K_1^T * \zeta) + K_0^T\zeta(0),$$

$$[A_K^T\Psi]^1(\tau) = \dot{\Psi}^1(\tau) + K_1^T(\tau)\mathcal{B}^T\Psi^1$$

$$= \frac{d}{d\tau}\left(\psi^1(\tau) + K_1^T * \zeta(\tau)\right) + K_1^T(\tau)\zeta(0)$$

$$= \dot{\psi}^1(\tau) + K_1^T * \dot{\zeta}(\tau), \ -h \leqslant \tau \leqslant 0.$$

We conclude that $T(\psi,\zeta) \in \operatorname{dom} A_K^T$ and $A_K^T T(\psi,\zeta) = T\tilde{A}_K^T(\psi,\zeta)$ for every $(\psi,\zeta) \in \operatorname{dom} \tilde{A}_K^T$. By Lemma 1.3.8, this implies

$$S_K^T(t)T = T\tilde{S}_K^T(t), \ t \geqslant 0. \tag{22}$$

Thus we have proved (21) for the case $y(t) \equiv 0$.

In order to prove the second part of formula (21), we define $Z(t) \in \mathbb{R}^{n\times m}$, $t \geqslant 0$, and $X(t) \in \mathbb{R}^{n\times m}$, $V(t) \in \mathbb{R}^{m\times m}$, $t \geqslant -h$, to be the unique solution of the homogeneous system Σ_K^T $(y(t) \equiv 0)$ corresponding to the initial condition

$$Z(0) = K_0^T, \ X(\tau) = K_1^T(\tau), \ V(\tau) = 0,$$

for $-h \leqslant \tau < 0$ (Theorem 1.2.3). Then, by (22), we have

$$(Z(t), X_t + K_t^T * V_t) = S_K^T(t)K^T, \ t \geqslant 0. \tag{23}$$

Moreover, we define

$$\begin{aligned}
z(t) &= -\int_0^t Z(t-s)y(s)ds \\
x(t) &= -\int_0^t X(t-s)y(s)ds, \ x(\tau) = 0, \\
v(t) &= -\int_0^t V(t-s)y(s)ds - y(t), \ v(\tau) = 0, \\
x(t,\tau) &= -\int_0^t \left[X_{t-s}(\tau) + K_1^T * V_{t-s}(\tau)\right]y(s)ds,
\end{aligned} \tag{24}$$

for $t \geqslant 0$ and $-h \leqslant \tau \leqslant 0$. Then we obtain

$$x(t,\tau) = -\int_0^t X(t-s+\tau)y(s)ds - \int_0^t \int_\tau^0 K_1^T(\tau-\sigma)V(t-s+\sigma)y(s)d\sigma ds$$

$$= x(t+\tau) - \int_{t+\tau}^t X(t-s+\tau)y(s)ds$$

$$- \int_\tau^0 K_1^T(\tau-\sigma) \int_0^{t+\sigma} V(t-s+\sigma)y(s)dsd\sigma$$

$$= x(t+\tau) + \int_\tau^0 K_1^T(\tau-\sigma)v(t+\sigma)d\sigma$$

and hence $x(t,\cdot) = x_t + K_1^T * v_t$, $t \geqslant 0$.
This implies

$$\dot{z}(t) = -\int_0^t \dot{Z}(t-s)y(s)ds - Z(0)y(t)$$

$$= -\int_0^t \int_{-h}^0 d\eta^T(\tau)\left[X_{t-s}(\tau) + K_1^T * V_{t-s}(\tau)\right]y(s)ds$$

$$- K_0^T \int_0^t V(t-s)y(s)ds - K_0^T y(t)$$

$$= L^T(x(t,\cdot)) + K_0^T v(t)$$

$$= L^T(x_t + K_1^T * v_t) + K_0^T v(t)$$

and by analogy

$$x(t) = z(t) + M^T(x_t + K_1^T * v_t),$$

$$v(t) = B^T(x_t + K_1^T * v_t) - y(t).$$

We conclude that the triple $z(t)$, $x(t)$, $v(t)$ satisfies Σ_K^T, which means $\hat{x}(t) = (z(t), x_t + K_1^T * v_t) = (z(t), x(t,\cdot))$. Hence, by (23) and (24), the following equation holds for every $\phi \in M^p$

$$\langle \hat{x}(t), \phi \rangle$$

$$= \langle (z(t), x(t,\cdot)), \phi \rangle$$

$$= - \int_0^t \left[Z(t-s)y(s)\right]^T \phi^0 ds$$

$$- \int_{-h}^0 \int_0^t \left[X_{t-s}(\tau)y(s) + K_1^T * V_{t-s}(\tau)y(s)\right]^T \phi^1(\tau)dsd\tau$$

$$= - \int_0^t \langle S_K^T(t-s)K^T y(s),\phi\rangle ds$$

$$= \langle - \int_0^t S_K^T(t-s)K^T y(s)ds,\phi\rangle. \quad \square$$

5.2.2 REMARKS

(i) If $B^T = 0$, then $S_K^T(t) = S^T(t)$ for every $t \geq 0$. In this case Theorem 5.2.1 leads to the following interesting interpretation of an (arbitrary, finite-dimensional) input operator $K^T : \mathbb{R}^m \to M^q$ for the semigroup $S^T(t)$.

Let $z \in W^{1,q}_{loc}([0,\infty);\mathbb{R}^n)$ *and* $x \in L^q_{loc}([-h,\infty);\mathbb{R}^n)$ *satisfy the equations*

$$\begin{aligned} \dot{z}(t) &= L^T(x_t + K_1^T * v_t) + K_0^T v(t), \\ x(t) &= z(t) + M^T(x_t + K_1^T * v_t), \end{aligned} \tag{25}$$

for some $v \in L^q_{loc}([-h,\infty);\mathbb{R}^m)$. *Then the evolution of the pair* $\hat{x}(t) = (z(t),x_t + K_1^T * v_t) \in M^q$ *is described by the variation-of-constants formula*

$$\hat{x}(t) = S^T(t)\hat{x}(0) + \int_0^t S^T(t-s)K^T v(s)ds. \tag{26}$$

(ii) In the case $K_1(\tau) \equiv 0$ equation (26) reduces to the 'classical' variation-of-constants formula (2.44).

(iii) For retarded systems with undelayed output variables Theorem 5.2.1 has been proved in Salamon [135, Theorem 4.2]. A preliminary version of this result for RFDEs with output delays in the state space C can be found in Salamon [133].

Note that formula (21) describes the mild solutions of the abstract evolution equation

$$d/dt\ \hat{x}(t) = A_K^T\ \hat{x}(t) - K^T y(t) \tag{27}$$

in the Banach space M^q. This is precisely the observer equation which was introduced in Section 1.3 (compare equations (1.34) and (1.35)). Hence we have to check the stability of the semigroup $S_K^T(t)$ on M^q.

On the level of delay equations, system Σ_K^T can be regarded as a concrete observer equation for system Σ^T. In fact, it is easy to see that, for any solution $z(t)$, $x(t)$ of Σ^T and any corresponding solution $\tilde{z}(t)$, $\tilde{x}(t)$, $v(t)$ of Σ_K^T, the 'error'

$$g(t) = \tilde{z}(t) - z(t), \; e(t) = \tilde{x}(t) - x(t),$$

together with $v(t)$ satisfies the homogeneous equation Σ_K^T (note that the variable $v(t)$ in Σ_K^T may be interpreted as the 'error of the output'). The evolution of this triple $(g(t), e_t, v_t)$ is described by the semigroup $\tilde{S}_K^T(t)$ on $M^q \times L^q$ which was introduced in the proof of Theorem 5.2.1. Therefore we will also analyze the stability behaviour of this semigroup.

To do this we need the following characterization of the spectrum of the operator A_K^T via the complex matrix function

$$\Delta_K^T(\lambda) = \begin{bmatrix} \Delta^T(\lambda) & -K_0^T - L^T(K_1^T * e^{\lambda\cdot}) - \lambda M^T(K_1^T * e^{\lambda\cdot}) \\ -B^T(e^{\lambda\cdot}) & I - B^T(K_1^T * e^{\lambda\cdot}) \end{bmatrix}. \tag{28}$$

5.2.3 <u>LEMMA</u>

(i) *The exponential growth of system* Σ_K^T *is given by*

$$\omega_K := \lim_{t\to\infty} t^{-1} \log \|S_K^T(t)\|_{L(M^q)}$$

$$= \lim_{t\to\infty} t^{-1} \log \|\tilde{S}_K^T(t)\|_{L(M^q \times L^q)}.$$

(ii) *Let* $\lambda \in \mathbb{C}$ *and* $\psi, \Psi \in M^q$ *be given. Then* $\psi \in \text{dom } A_K^T$ *and* $(\lambda I - A_K^T)\psi = \Psi$ *if and only if*

$$\psi^1(\tau) = e^{\lambda\tau}\psi^1(0) + \int_\tau^0 e^{\lambda(\tau-\sigma)}(\Psi^1(\sigma) + K_1^T(\sigma)B^T\psi^1)d\sigma, \tag{29.1}$$

$$\psi^0 = \psi^1(0) - M^T\psi^1, \tag{29.2}$$

$$\Delta_K^T(\lambda)\begin{pmatrix} \psi^1(0) \\ B^T\psi^1 \end{pmatrix} = \begin{pmatrix} \Psi^0 + L^T(e^{\lambda\cdot} * \Psi^1) + \lambda M^T(e^{\lambda\cdot} * \Psi^1) \\ B^T(e^{\lambda\cdot} * \Psi^1) \end{pmatrix}. \tag{29.3}$$

(iii) *The resolvent operator* $(\lambda I - A_K^T)^{-1}$ *is compact for every* $\lambda \notin \sigma(A)$.

(iv) $\sigma(A_K^T) = P\sigma(A_K^T) = \sigma(\tilde{A}_K^T) = \{\lambda \in \mathbb{C} | \det \Delta_K^T(\lambda) = 0\}$.

Proof. (i) Note that the operator $T : M^q \times L^q \to M^q$, introduced in the proof of Theorem 5.2.1, is surjective. Hence, by (22), the exponential growth ω_K of $S_K^T(t)$ is not larger than the exponential growth of $\tilde{S}_K^T(t)$. On the other hand, let $\omega > \omega_k$ and let $z(t)$, $x(t)$, $v(t)$ be any solution of the homogeneous system Σ_K^T. Then it follows again from (22) that the functions

$$|z(t)|e^{-\omega t}, \quad \|x_t + K_1^T * v_t\|_q \, e^{-\omega t}$$

tend to zero if t goes to infinity. Hence, by the last two equations in Σ_K^T, the function $t \to \|(z(t), x_t, v_t)\| \, e^{-\omega t}$ is bounded on $[0,\infty)$. This proves that the exponential growth of the semigroup $\tilde{S}_K^T(t)$ is less than or equal to ω.

(ii) Recall that the operator A_K^T is given by (19). Hence $\psi \in \operatorname{dom} A_K^T$ and $(\lambda I - A_K^T)\psi = \Psi$ if and only if $\psi \in W^{1,q}$, $\psi^0 = \psi^1(0) - M^T\psi^1$, and

$$\lambda\psi^0 - L^T\psi^1 - K_0^T B^T \psi^1 = \Psi^0, \tag{30.1}$$

$$\lambda\psi^1(\tau) - \dot{\psi}^1(\tau) - K_1^T(\tau)B^T\psi^1 = \Psi^1(\tau), \; -h \leqslant \tau \leqslant 0. \tag{30.2}$$

Note that (29.1) is equivalent to (30.2). If (29.1) is satisfied, then (30.1) is equivalent to

$$\begin{aligned}\Psi^0 + K_0^T B^T \psi^1 &= \lambda\psi^1(0) - L^T\psi^1 - \lambda M^T\psi^1 \\ &= \Delta(\lambda)\psi^1(0) - L^T(e^{\lambda\cdot} * \Psi^1) - \lambda M^T(e^{\lambda\cdot} * \Psi^1) \\ &\quad - L^T(K_1^T * e^{\lambda\cdot})B^T\psi^1 - \lambda M^T(K_1^T * e^{\lambda\cdot})B^T\psi^1.\end{aligned}$$

Also, $B^T\psi^1$ is given by

$$B^T\psi^1 = B^T(e^{\lambda\cdot})\psi^1(0) + B^T(e^{\lambda\cdot} * \Psi^1) + B^T(K_1^T * e^{\lambda\cdot})B^T\psi^1.$$

These two equations are equivalent to (29.3).

By Theorem 1.2.7, the spectrum of the operator $\tilde{A}_K^T$ (i.e., of system Σ_K^T) is characterized by the complex matrix function

$$\tilde{\Delta}_K^T(\lambda) = \begin{bmatrix} \lambda I & -L^T(e^{\lambda\cdot}) & -K_0^T - L^T(K_1^T * e^{\lambda\cdot}) \\ -I & I - M^T(e^{\lambda\cdot}) & -M^T(K_1^T * e^{\lambda\cdot}) \\ 0 & -B^T(e^{\lambda\cdot}) & -B^T(K_1^T * e^{\lambda\cdot}) \end{bmatrix} .$$

Some elementary operations show that this matrix becomes nonsingular if and only if det $\Delta_K^T(\lambda) \neq 0$. Now (iii) and (iv) follow easily from (ii). □

In the following we will assume that M is given by (3). Then the next result shows that the stability of the difference equation Σ_0^T is a necessary condition for the stability of the closed loop system Σ_K^T.

5.2.4 LEMMA. *Let $\alpha < \beta$ be given. Then the following statements are equivalent.*

(i) *There exists some* $\lambda_0 \in \mathbb{C}$ *such that* $\alpha < \operatorname{Re}\lambda < \beta$ *and* $\det \Delta_0^T(\lambda_0) = 0$.

(ii) *There exists some* $\varepsilon > 0$ *and a sequence* $\lambda_k \in \mathbb{C}$ *such that* $|\operatorname{Im}\lambda_k|$ *tends to infinity,* $\alpha + \varepsilon \leq \operatorname{Re}\lambda_k \leq \beta - \varepsilon$, *and* $\det \Delta_K^T(\lambda_k) = 0$ *for every* $k \in \mathbb{N}$.

Proof. Note that λ^{-n} det $\Delta_K^T(\lambda)$ is the determinant of the matrix

$$\begin{bmatrix} \Delta_0^T(\lambda)-\lambda^{-1}L^T(e^{\lambda\cdot})-\int_{-h}^{0} A_{-\infty}^T(\tau)e^{\lambda\tau}d\tau & -\lambda^{-1}K_0^T-\lambda^{-1}L^T(K_1^T*e^{\lambda\cdot})-M^T(K_1^T*e^{\lambda\cdot}) \\ -B^T(e^{\lambda\cdot}) & I-B^T(K_1^T*e^{\lambda\cdot}) \end{bmatrix} .$$

This implies that the limit

$$\lim_{|\operatorname{Im}\lambda| \to \infty} |\lambda^{-n} \det \Delta_K^T(\lambda) - \det \Delta_0^T(\lambda)| = 0$$

exists uniformly for $\alpha \leq \operatorname{Re}\lambda \leq \beta$. The rest of the proof is precisely the same as in Lemma 5.1.2. □

We are now going to show that the spectrum of the generator A_K^T determines the exponential growth of the semigroup $S_K^T(t)$. For this sake we need the concept of a *Fredholm operator* (Kato [68, p. 230]). A bounded linear operator $T : X \to X$ on a Banach space X is said to be a Fredholm operator if

ran T is closed,
dim ker T $< \infty$,
codim ran T $< \infty$.

The following result can be found in Kato [68, p. 238, Theorem 5.26].

5.2.5 THEOREM. *Let* X *be a Banach space,* $T \in L(X)$, *and* $K \in L(X)$ *compact. Then* T *is a Fredholm operator if and only if* $T + K$ *is a Fredholm operator.*

Let us now introduce the semigroup $S_0^T(t) : M^q \to M^q$ by defining $S_0^T(t)\psi = (0, x_t)$ for $\psi \in M^q$ where

$$x(t) = \sum_{j=1}^{\infty} A_{-j}^T \, x(t-h_j), \; t \geqslant 0,$$

$$x(\tau) = \psi^1(\tau), \; -h \leqslant \tau < 0.$$

The following result has been proved by Henry [50, Theorem 3.2 and Lemma 4.1] in the context of the state space C. The same arguments apply to the product space situation.

5.2.6 LEMMA.

(i) *Let* $s \in \sigma(S_0^T(t))$, $s \neq 0$. *Then*

$$|s| \in \mathrm{cl}(\{e^{\mathrm{Re}\,\lambda t} \,|\, \det \Delta_0^T(\lambda) = 0\}).$$

(ii) *The operator* $S^T(t) - S_0^T(t)$ *is compact.*

The most difficult part in the proof of this result is statement (i). It is the main step towards our desired 'spectrum determined growth' condition[1] for the semigroup $S_K^T(t)$.

5.2.7 THEOREM. *Let* M *be given by* (3). *Then*

$$\omega_K = \lim_{t\to\infty} t^{-1} \log \|S_K^T(t)\|_{L(M^q)}$$

$$= \sup \{\mathrm{Re}\,\lambda \,|\, \det \Delta_K^T(\lambda) = 0\}.$$

Proof. Suppose that $\omega_K > \sup \{\mathrm{Re}\,\lambda \,|\, \det \Delta_K^T(\lambda) = 0\}$ and note that the spectral radius of the operator $S_K^T(t)$ is given by $e^{\omega_K t}$ (Zabczyk [149, Lemma 1]). Then there exists some $s \in \sigma(S_K^T(t))$ such that

---

(1) This notion has been introduced by Triggiani-Pritchard [144].

$$|s| > \sup \{e^{\mathrm{Re}\,\lambda t} | \det \Delta_K^T(\lambda) = 0\}$$

$$\geq \sup \{e^{\mathrm{Re}\,\lambda t} | \det \Delta_0^T(\lambda) = 0\}$$

(Lemma 5.2.4). Now let $\lambda \in \mathbb{C}$ satisfy $e^{\lambda t} = s$. Then $\det \Delta_K^T(\lambda) \neq 0$ and hence λ is neither in the point spectrum of A_K^T nor in the point spectrum of A_K^{T*} (Lemma 5.2.3). By Hille-Phillips [53, Theorem 16.7.2], this implies

$$s = e^{\lambda t} \notin P\sigma(S_K^T(t)) \cup P\sigma(S_K^{T*}(t))$$

$$= P\sigma(S_K^T(t)) \cup R\sigma(S_K^T(t)).$$

We conclude that $s \in C\sigma(S_K^T(t))$ and hence $sI - S_K^T(t)$ is not a Fredholm operator.

Now we make use of the fact that $S_K^T(t) - S^T(t)$ is a compact operator (Corollary 1.3.11). By Lemma 5.2.6 (ii), this implies that the operator $S_K^T(t) ; S_0^T(t)$ is also compact. Applying Theorem 5.2.5, we obtain that $sI - S_0^T(t)$ is not a Fredholm operator. This is a contradiction to Lemma 5.2.6 (i). □

Let us now discuss briefly the properties of the observer semigroup for system Ω^T ($\Gamma = 0$). For this purpose we have to assume that ran $K^T \subset$ ran ι^T, i.e.

$$\mathcal{K}^T(\cdot) := K_1^T(\cdot) \in W^{1,q}([-h,0];\mathbb{R}^{n\times m}),$$

$$K_0^T = \mathcal{K}^T(0) - M^T\mathcal{K}^T.$$

This means that

$$K^T = \iota^T \mathcal{K}^T \tag{31}$$

where we have identified the *function* $\mathcal{K}^T$ with the *operator* $\mathcal{K}^T : \mathbb{R}^m \to W^{1,q}$ which maps $y \in \mathbb{R}^m$ into $\mathcal{K}^T(\cdot)y \in W^{1,q}$. Now the observer semigroup $S_{\mathcal{K}}^T(t) : W^{1,q} \to W^{1,q}$ of system Ω^T is generated by the boundedly perturbed operator $A_{\mathcal{K}}^T = A^T + \mathcal{K}^T\mathcal{B}^T$. By Theorem 1.3.9 (iv), this semigroup satisfies

$$\iota^T S_{\mathcal{K}}^T(t) = S_K^T(t)\iota^T, \; t \geq 0. \tag{32}$$

Hence the operator $(sI - A_K^T)\iota^T : W^{1,q} \to M^q$, $s \notin \sigma(A_K^T)$, is a similarity action between the semigroups $S_K^T(t)$ and $S_K^T(t)$ (Lemma 1.3.2 (iii)). This implies that the semigroup $S_K^T(t)$ has properties analogous to those of $S_K^T(t)$. The main facts are summarized in the corollary below.

5.2.8 <u>COROLLARY</u> *Suppose that* M *is given by* (3) *and* K^T *by* (31).

(i) *Let* $y \in L^q_{loc}([0,\infty);\mathbb{R}^m)$ *be given and let the pair* $x \in W^{1,q}_{loc}([-h,\infty);\mathbb{R}^m)$, $v \in L^q_{loc}([-h,\infty);\mathbb{R}^m)$ *satisfy the equations*

$$\Omega_K^T \qquad \boxed{\begin{aligned} \dot{x}(t) &= L^T(x_t + K^T * v_t) + K^T(0)v(t) \\ &\qquad + M^T(\dot{x}_t + \dot{K}^T * v_t - K^T(0)v_t) \\ v(t) &= B^T(x_t + K^T * v_t) - y(t) \end{aligned}}$$

for $t \geq 0$. *Then*

$$\hat{x}(t) = x_t + K^T * v_t \in W^{1,q} \tag{33}$$

is given by the variation-of-constants formula

$$\hat{x}(t) = S_K^T(t)\hat{x}(0) - \int_0^t S_K^T(t-s)K^T y(s)ds, \ t \geq 0. \tag{34}$$

(ii) *The generator* A_K *has a pure point spectrum*

$$\sigma(A_K^T) = P\sigma(A_K^T) = \{\lambda \in \mathbb{C} \mid \det \Delta_K^T(\lambda) = 0\}$$

and the resolvent operator $(\lambda I - A_K^T)^{-1}$ *is compact for* $\lambda \notin \sigma(A_K^T)$.

(iii) *The exponential growth of system* Ω_K^T *is given by*

$$\omega_K = \lim_{t\to\infty} t^{-1} \log \|S_K^T(t)\|_{L(W^{1,q})}$$

$$= \sup \{\text{Re } \lambda \mid \lambda \in \sigma(A_K^T)\}.$$

<u>Proof</u>. (i) Let the pair $x(t)$, $v(t)$, $t \geq -h$, satisfy Ω_K^T and define

$x(t) := x(t)$, $v(t) := v(t)$ for $t \geq -h$ as well as $z(t) := x(t) - M^T(x_t + K^T * v_t)$ for $t \geq 0$. Moreover, let K^T be given by (31). Then the following equation holds

$$\dot{z}(t) = \dot{x}(t) - M^T\dot{x}_t - \int_{-h}^{0} d\mu^T(\tau)\left(\frac{d}{dt}\int_{t+\tau}^{t} K^T(t+\tau-s)v(s)ds\right)$$

$$= \dot{x}(t) - M^T\left(\dot{x}_t + K^T v(t) - K^T(0)v_t + \dot{K}^T * v_t\right)$$

$$= L^T\left(x_t + K^T * v_t\right) + \left(K^T(0) - M^T K^T\right)v(t)$$

$$= L^T\left(x_t + K_1^T * v_t\right) + K_0^T v(t).$$

Hence the triple $z(t)$, $x(t)$, $v(t)$ satisfies Σ_K^T. By Theorem 5.2.1, this implies that $\hat{x}(t) = (z(t), x_t + K_1^T * v_t) = \iota^T\hat{x}(t) \in M^q$ is given by

$$\iota^T\hat{x}(t) = S_K^T(t)\iota^T\hat{x}(0) - \int_0^t S_K^T(t-s)\iota^T K^T y(s)ds$$

$$= \iota^T\left(S_K^T(t)\hat{x}(0) - \int_0^t S_K^T(t-s)K^T y(s)ds\right).$$

This proves (34).

Statements (ii) and (iii) follow directly from the similarity of the semigroups $S_K^T(t)$ and $S_K^T(t)$ together with Lemma 5.2.3 and Theorem 5.2.7. □

THE FEEDBACK SEMIGROUP

Let us begin with the discussion of system $\tilde{\Omega}$.

We have seen that the state $\hat{x}(t) = (x(t), x^t) \in M^p$ of $\tilde{\Omega}$ at time $t \geq 0$, corresponding to some input $u \in L^p_{loc}([0,\infty);\mathbb{R}^m)$ and some initial state $f \in M^p$, is described by the variation-of-constants formula

$$\hat{x}(t) = S^{T*}(t)f + \iota^{T*^{-1}} \int_0^t S^{T*}(t-s)B^{T*}u(s)ds. \tag{35}$$

(Theorem 2.3.6). Moreover the input operator $B^{T*} : R^m \to W^{-1,p}$ satisfies the hypothesis (H2) of Section 1.3 (see Remark 2.3.7). This implies that the state $\hat{x}(t) \in M^p$ of system $\tilde{\Omega}$ is the unique solution of the Cauchy problem

$$d/dt\ \iota^{T*}\hat{x}(t) = A^{T*}\ \iota^{T*}\hat{x}(t) + B^{T*}\ u(t),$$
$$x(0) = f \in M^p, \tag{36}$$

in the sense of Definition 1.3.3 (see page 77).

We want to apply Theorem 1.3.7 in order to obtain a feedback semigroup for system $\tilde{\Omega}$ in the state space M^p. Therefore we allow control laws of the general form

$$u(t) = K^{T*}\hat{x}(t) = K_0 x(t) + \int_{-h}^{0} K_1(\sigma)x^t(\sigma)d\sigma. \tag{37}$$

Then, by Theorem 1.3.7 (i), there exists a unique C_0-semigroup $S_K^{T*}(t):M^p \to M^p$ such that the following equation holds for every $f \in M^p$ and every $t \geq 0$

$$\iota^{T*}S_K^{T*}(t)f = \iota^{T*}S^{T*}(t)f + \int_0^t S^{T*}(t-s)B^{T*}K^{T*}S_K^{T*}(s)f ds. \tag{38}$$

This equation can be obtained by inserting (37) into (35) and replacing $\hat{x}(t)$ by $S_K^{T*}(t)f$. Hence $S_K^{T*}(t)$ is in fact a feedback semigroup for system $\tilde{\Omega}$. Moreover, it follows from the equations (38) and (18) that this feedback semigroup is precisely the adjoint of the observer semigroup $S_K^T(t)$ for system Σ^T.

The infinitesimal generator of $S_K^{T*}(t)$ is given by

$$\text{dom}\ A_K^{T*} = \left\{ f \in M^p \,\middle|\, A^{T*}\ \iota^{T*}f + B^{T*}K^{T*}f \in \text{ran}\ \iota^{T*} \right\}$$
$$\iota^{T*}A_K^{T*}f = A^{T*}\ \iota^{T*}f + B^{T*}K^{T*}f \tag{39}$$

(Theorem 1.3.7 (iii)). Its spectrum can be characterized via the complex matrix function

$$\Delta_K(\lambda) = \begin{bmatrix} \Delta(\lambda) & -\ B(e^{\lambda\cdot}) \\ -\langle K^T, Fe^{\lambda\cdot}\rangle & I - \langle K^T, Ee^{\lambda\cdot}\rangle \end{bmatrix} \tag{40}$$

Note that this is the transposed of the matrix $\Delta_K^T(\lambda)$ which is defined by (28).

5.2.9 THEOREM

(i) *Let* $x \in W^{1,p}_{loc}([-h,\infty);\mathbb{R}^n)$ *and* $u \in L^p_{loc}([-h,\infty);\mathbb{R}^m)$ *satisfy the equations*

$$\Omega_K \quad \boxed{\begin{aligned} \dot{x}(t) &= Lx_t + M\dot{x}_t + Bu_t \\ u(t) &= K_0 x(t) + \int_{-h}^{0}\int_{\tau}^{0} K_1(\tau-\sigma)d\eta(\tau)x(t+\sigma)d\sigma \\ &\quad + \int_{-h}^{0}\int_{\tau}^{0} K_1(\tau-\sigma)d\mu(\tau)\dot{x}(t+\sigma)d\sigma \\ &\quad + \int_{-h}^{0}\int_{\tau}^{0} K_1(\tau-\sigma)d\beta(\tau)u(t+\sigma)d\sigma \end{aligned}}$$

and define

$$\hat{x}(t) = Fx_t + Eu_t \in M^p. \tag{41}$$

Then $u(t) = K^{T*}\hat{x}(t)$ *and*

$$\hat{x}(t) = S_K^{T*}(t)\hat{x}(0), \; t \geqslant 0. \tag{42}$$

(ii) *The generator* A_K^{T*} *has a pure point spectrum*

$$\sigma(A_K^{T*}) = P\sigma(A_K^{T*}) = \{\lambda \in \mathbb{C} | \det \Delta_K(\lambda) = 0\}$$

and the resolvent operator $(\lambda I - A_K^{T*})^{-1}$ *is compact.*

(iii) *If* M *is defined by* (3), *then the exponential growth of system* Ω_K *is given by*

$$\begin{aligned} \omega_K &= \lim_{t\to\infty} t^{-1} \log \|S_K^{T*}(t)\|_{L(M^p)} \\ &= \sup \{\text{Re } \lambda | \lambda \in \sigma(A_K^{T*})\}. \end{aligned}$$

(iv) *Let* $f, g \in M^p$ *be given. Then* $g \in \text{dom } A_K^{T*}$ *and* $A_K^{T*}g = f$ *if and only if the following equations hold*

$$-\eta(-h)g^0 - \beta(-h)K^{T*}g = \Big(I + \mu(-h)\Big)f^0 + \int_{-h}^{0} f^1(\tau)d\tau,$$

$$g^1(\sigma) - \eta(\sigma)g^0 - \beta(\sigma)K^{T*}g = \Big(I + \mu(\sigma)\Big)f^0 + \int_{\sigma}^{0} f^1(\tau)d\tau, \; -h \leqslant \tau \leqslant 0.$$

Proof. (i) If $x(t)$ and $u(t)$ satisfy Ω_K and if $\hat{x}(t)$ is defined by (41), then it follows from the definition of the operators F (page 55) and E (page 78) that $u(t) = K^{T*}\hat{x}(t)$. Moreover, the first equation in Ω_K implies

$$\hat{x}(t) = S^{T*}(t)\hat{x}(0) + \iota^{T*^{-1}} \int_0^t S^{T*}(t-s)B^{T*}u(s)ds$$

(Corollary 2.3.8). Hence (42) follows from the definition of the semigroup $S_K^{T*}(t)$.

Now we prove that the semigroup $S_K^{T*}(t)$ is stable with exponential decay rate ω if and only if system Ω_K has the same property which means that the functions

$$|x(t)|e^{-\omega t}, \ |u(t)|e^{-\omega t}, \ t \geq 0,$$

are bounded for every solution of Ω_K. If the semigroup $S_K^{T*}(t)$ is stable, then the stability of Ω_K follows from statement (i). Conversely, let system Ω_K be stable. Then it follows again from (i) that the function

$$\|S_K^{T*}(t)\phi\|e^{-\omega t}, \ t \geq 0, \tag{43}$$

is bounded for every $\phi \in$ ran $[F\ E]$. Now it follows from (38) that

$$\text{ran } S_K^{T*}(t) \subset \text{ran } [F\ E] \ \forall\ t \geq h \tag{44}$$

(see Theorem 2.2.3 and Corollary 2.3.8). Hence the function (43) is bounded for every $\phi \in M^p$ and the stability of $S_K^{T*}(t)$ is a consequence of the uniform boundedness theorem.

Statement (ii) and the remainder of (iii) follow from Lemma 5.2.3 and Theorem 5.2.7 by duality.

(iv) It follows from (19) that $g \in$ dom A_K^{T*} and $A_K^{T*}g = f$ if and only if $\langle\iota^T\psi, f\rangle = \langle A_K^T\iota^T\psi, g\rangle$ for every $\psi \in W^{1,q}$. Hence statement (iv) is a consequence of

$$\begin{aligned}
&\langle\iota^T\psi, f\rangle \\
&= (\psi(0) - M^T\psi)^T f^0 + \int_{-h}^0 \psi^T(\tau)f^1(\tau)d\tau \\
&= \psi^T(-h)\Big((I + \mu(-h))f^0 + \int_{-h}^0 f^1(\tau)d\tau\Big) + \int_{-h}^0 \dot{\psi}^T(\sigma)\Big((I+\mu(\sigma))f^0 + \int_\sigma^0 f^1(\tau)d\tau\Big)d\sigma
\end{aligned}$$

and

$$\langle A_{K\iota}^{T}{}^{T}\psi, g\rangle$$

$$= (L^T\psi + K_0^T B^T\psi)^T g^0 + \int_{-h}^{0}\left(\dot{\psi}(\sigma) + K_1^T(\sigma)B^T\psi\right)^T g^1(\sigma)d\sigma$$

$$= \int_{-h}^{0}\psi^T(\tau)d\eta(\tau)g^0 + \int_{-h}^{0}\psi^T(\tau)d\beta(\tau)K^{T*}g + \int_{-h}^{0}\dot{\psi}^T(\sigma)g^1(\sigma)d\sigma$$

$$= \psi^T(-h)\left(-\eta(-h)g^0 - \beta(-h)K^{T*}g\right) + \int_{-h}^{0}\dot{\psi}^T(\sigma)\left(g^1(\sigma) - \eta(\sigma)g^0 - \beta(\sigma)K^{T*}g\right)d\sigma. \quad \square$$

5.2.10 REMARKS

(i) System Ω_K admits a unique solution for every initial condition of the form (2.47).

In fact, the introduction of the new variable $z(t) := \dot{x}(t)$ in Ω_K leads to the following equivalent system of the form (1.13)

$$\dot{x}(t) = z(t), \tag{45.1}$$

$$z(t) = Lx_t + Mz_t + Bu_t, \tag{45.2}$$

$$u(t) = K_0x(t) + \int_{-h}^{0}\int_{\tau}^{0} K_1(\tau-\sigma)d\eta(\tau)x(t+\sigma)d\sigma$$

$$+ \int_{-h}^{0}\int_{\tau}^{0} K_1(\tau-\sigma)d\mu(\tau)z(t+\sigma)d\sigma \tag{45.3}$$

$$+ \int_{-h}^{0}\int_{\tau}^{0} K_1(\tau-\sigma)d\beta(\tau)u(t+\sigma)d\sigma.$$

Hence the above statement follows from Theorem 1.2.3.

(ii) Note that system (45) can be obtained from Σ_K^T by transposition of matrices. In particular, both systems have the same spectrum, characterized by the complex matrix function $\tilde{\Delta}_K^T(\lambda)$ which is defined on page 175 .

At the end of this section we consider the feedback semigroup for system $\tilde{\Sigma}$.

Recall that the state $\hat{x}(t) = \pi(w(t), w^t, x^t) \in W^{-1,p}$ of $\tilde{\Sigma}$ at time $t \geqslant 0$, corresponding to some input $u \in L^p([0,\infty);\mathbb{R}^m)$, is given by the variation-of-constants formula

$$\hat{x}(t) = S^{T*}(t)\hat{x}(0) + \int_0^t S^{T*}(t-s)B^{T*}u(s)ds \tag{46}$$

(Theorem 2.2.6). This means that $\hat{x}(t)$ is a mild solution of the Cauchy problem

$$d/dt\ \hat{x}(t) = A^{T*}\hat{x}(t) + B^{T*}u(t) \tag{47}$$

(see page 77).

In order to transform (37) into a control law for (47), we have to assume that $K^{T*} : M^p \to \mathbb{R}^m$ can be extended to a bounded linear functional on $W^{-1,p}$. This means that

$$K^{T*} = K^{T*}\iota^{T*} \tag{48}$$

for some $K^{T*} \in L(W^{-1,p},\mathbb{R}^m)$ (compare equation (31)). If such a factorization is possible, then the operator K^{T*} can be represented by the matrix function $K(\cdot) = K_1(\cdot) \in W^{1,q}([-h,0];\mathbb{R}^{m\times n})$ in the following way

$$K^{T*}\pi f = K(0)f^0 + \int_{-h}^0 \left(K(\tau)f^1(\tau) + \dot{K}(\tau)f^2(\tau)\right)d\tau,\ f \in M^p. \tag{49}$$

Applying the control law

$$u(t) = K^{T*}\hat{x}(t) \tag{50}$$

to the Cauchy problem (47), we obtain the perturbed semigroup $S_K^{T*}(t) : W^{-1,p} \to W^{-1,p}$ which is generated by $A_K^{T*} = A^{T*} + B^{T*}K^{T*}$. This is the adjoint of the observer semigroup $S_K^T(t)$ for system Ω^T. Hence the feedback semigroup $S_K^{T*}(t)$ of systems $\tilde{\Sigma}$ can be regarded as an extension of the feedback semigroup $S_K^{T*}(t)$ of system $\tilde{\Omega}$ to the state space $W^{-1,p}$ - if K^{T*} is given by (48). This fact is formalized in the following equation

$$S_K^{T*}(t)\iota^{T*} = \iota^{T*}S_K^{T*}(t),\ t \geqslant 0, \tag{51}$$

(compare (32)).

The main properties of the semigroup $S_K^{T*}(t)$ are summarized in the theorem below. The proof will be omitted since it is analogous to that of Theorem 5.2.9.

5.2.11 THEOREM

(i) *Let the complex matrix* $\Delta_K(\lambda)$ *be given by* (40) *where* K^{T*} *satisfies* (48). *Then*

$$\sigma(A_K^{T*}) = P\sigma(A_K^{T*}) = \{\lambda \in \mathbb{C} | \det \Delta_K(\lambda) = 0\}$$

and the resolvent operator $(\lambda I - A^{T*})^{-1}$ *is compact for* $\lambda \notin \sigma(A_K^{T*})$.

(ii) *Let* $w \in W_{loc}^{1,p}([0,\infty);\mathbb{R}^n)$, $x \in L_{loc}^p([-h,\infty);\mathbb{R}^n)$, *and* $u \in L_{loc}^p([-h,\infty);\mathbb{R}^m)$ *satisfy the equations*

$$\Sigma_K \qquad \boxed{\begin{aligned} \dot{w}(t) &= Lx_t + Bu_t \\ x(t) &= w(t) + Mx_t \\ u(t) &= K(0)w(t) + \int_{-h}^{0}\int_{\tau}^{0} K(\tau-\sigma)d\mu(\tau)x(t+\sigma)d\sigma \\ &\quad + \int_{-h}^{0}\int_{\tau}^{0} \dot{K}(\tau-\sigma)d\mu(\tau)x(t+\sigma)d\sigma \\ &\quad + \int_{-h}^{0}\int_{\tau}^{0} K(\tau-\sigma)d\beta(\tau)u(t+\sigma)d\sigma \end{aligned}}$$

and define

$$\hat{x}(t) = F(w(t),x_t) + Eu_t \in W^{-1,p}. \tag{52}$$

Then $u(t) = K^{T*}\hat{x}(t)$ *and*

$$\hat{x}(t) = S_K^{T*}(t)\hat{x}(0), \quad t \geq 0. \tag{53}$$

(iii) *If* M *is defined by* (3), *then the exponential growth of system* Σ_K *is given by*

$$\begin{aligned} \omega_K &= \lim_{t\to\infty} t^{-1} \log \|S_K^{T*}(t)\|_{L(W^{-1,p})} \\ &= \sup \{\mathrm{Re}\ \lambda | \lambda \in \sigma(A_K^{T*})\}. \end{aligned}$$

5.3 FINITE POLE-SHIFTING

In the preceding section we have seen that the exponential growth ω_K of the closed loop systems Σ_K^T, Ω_K^T, Ω_K, Σ_K as well as of the closed loop semigroups $S_K^T(t)$, $\mathcal{S}_K^T(t)$, $S_K^{T*}(t)$, $\mathcal{S}_K^{T*}(t)$ is determined by the complex matrix function $\Delta_K(\lambda)$ if equation (31), respectively (48), is satisfied. The remaining problem is the following.

Given $\omega \leq 0$, find some function $K(\cdot) \in W^{1,q}([-h,0];\mathbb{R}^{m\times n})$, respectively some pair

$$K_0 \in \mathbb{R}^{m\times n}, \ K_1(\cdot) \in W^{1,q}([-h,0];\mathbb{R}^{m\times n}),$$

(satisfying equation (31)) such that $\omega_K < \omega$. This means that all zeros of $\det \Delta_K(\lambda)$ are contained in some given left halfplane $\{\lambda \in \mathbb{C} \mid \mathrm{Re}\,\lambda \leq \omega-\varepsilon\}$, $\varepsilon > 0$. By Lemma 5.2.4, this requires the condition

$$\det \Delta_0(\lambda) = 0 \Rightarrow \mathrm{Re}\,\lambda \leq \omega - \varepsilon. \tag{54}$$

If (54) is satisfied, then

$$\Lambda = \{\lambda \in \mathbb{C} \mid \det \Delta(\lambda) = 0, \ \mathrm{Re}\,\lambda \geq \omega\}$$

is a finite symmetric subset of the complex plane (Lemma 5.1.2). For this set we introduce the real generalized eigenspaces $X_\Lambda \subset W^{1,p}$, $X_\Lambda^T \subset W^{1,q}$ and the complementary subspaces $X^\Lambda \subset W^{1,p}$, $X^{\Lambda^T} \subset W^{1,q}$ associated with the operators A and A^T. Then $\dim X_\Lambda = \dim X_\Lambda^T =: N$ (Remark 2.4.5) and there exist bases

$$\Phi = [\phi_1 \ldots \phi_N] \in W^{1,p}([-h,0];\mathbb{R}^{n\times N}),$$

$$\Psi = [\psi_1 \ldots \psi_N] \in W^{1,q}([-h,0];\mathbb{R}^{n\times N})$$

of X_Λ and X_Λ^T such that

$$\langle \Psi, F\iota\Phi \rangle = \langle \iota^T\Psi, \mathcal{F}\Phi \rangle = I \tag{55}$$

(compare equation (2.62)). Under this condition the real $N \times N$-matrix A_Λ, defined by

$$A\Phi = \Phi A_\Lambda, \tag{56}$$

has the following properties

$$A^T\Psi = \Psi A_\Lambda^T, \tag{57.1}$$

$$\Psi(\tau) = \Psi(0)e^{A_\Lambda^T \tau}, \ -h \leqslant \tau \leqslant 0. \tag{57.2}$$

$$\sigma(A_\Lambda) = \Lambda, \tag{57.3}$$

(Proposition 2.4.7). Now let us define the input matrix

$$B_\Lambda = \int_{-h}^{0} \Psi^T(\tau)d\beta(\tau) \in \mathbb{R}^{N\times m}. \tag{58}$$

Then the lemma below is a consequence of Proposition 4.1.2 and the well-known finite-dimensional pole-shifting result.

5.3.1 LEMMA. *The following statements are equivalent.*

(i) *The matrix pair* A_Λ, B_Λ *is controllable.*

(ii) *For every symmetric set* Λ' *of* N *complex numbers there exists some real* $m\times N$*-matrix* K_Λ *such that* Λ' *is the spectrum of* $A_\Lambda + B_\Lambda K_\Lambda$.

(iii) *For every* $\lambda \in \Lambda$ *the following equation holds*

$$\text{rank}\,[\Delta(\lambda), B(e^{\lambda\cdot})] = n. \tag{59}$$

Given a matrix K_Λ as in the above lemma, we define

$$K(\tau) = K_\Lambda \Psi^T(\tau), \ -h \leqslant \tau \leqslant 0. \tag{60}$$

Then equation (31) implies that K_0 and $K_1(\cdot)$ are given by

$$K_1(\tau) = K_\Lambda \Psi^T(\tau), \ -h \leqslant \tau \leqslant 0, \tag{61.1}$$

$$K_0 = K_\Lambda \Psi^T(0) - K_\Lambda \int_{-h}^{0} \Psi^T(\tau)d\mu(\tau). \tag{61.2}$$

In this case the zeros of det $\Delta_K(\lambda)$ can be determined explicitly. This is done in the next theorem which generalizes Pandolfi's finite pole-shifting result [126, Theorem 3.1] to NFDEs with input delays.

5.3.2 THEOREM. *Let* K_0, $K_1(\cdot)$ *be given by* (61) *and* $\Delta_K(\lambda)$ *by* (40). *Then*

$$\{\lambda \in \mathbb{C} | \det \Delta_K(\lambda) = 0\}$$

$$= \{\lambda \in \mathbb{C} | \det \Delta(\lambda) = 0,\ \lambda \notin \Lambda\} \ \cup\ \sigma(A_\Lambda + B_\Lambda K_\Lambda).$$

Proof. First recall that

$$\Delta_K(\lambda) \begin{bmatrix} \Delta(\lambda) & - B(e^{\lambda\cdot}) \\ -K_\Lambda \langle \iota^T\Psi, Fe^{\lambda\cdot} \rangle & I - K_\Lambda \langle \iota^T\Psi, Ee^{\lambda\cdot}\rangle \end{bmatrix}. \tag{62}$$

Moreover, the following equation holds for every $\lambda \in \mathbb{C}$

$$(\lambda I - A_\Lambda) \langle \iota^T\Psi, Ee^{\lambda\cdot}\rangle$$

$$= (\lambda I - A_\Lambda) \int_{-h}^{0} \int_{\tau}^{0} \Psi^T(\tau-\sigma)e^{\lambda\sigma} d\sigma d\beta(\tau)$$

$$= \int_{-h}^{0} \left(\int_{\tau}^{0} (\lambda I - A_\Lambda)\, e^{(\lambda I - A_\Lambda)\sigma} d\sigma\right) e^{A_\Lambda \tau} \Psi^T(0) d\beta(\tau)$$

$$= \int_{-h}^{0} \left(I - e^{(\lambda I - A_\Lambda)\tau}\right) e^{A_\Lambda \tau} \Psi^T(0) d\beta(\tau) \tag{63}$$

$$= \int_{-h}^{0} \Psi^T(\tau) d\beta(\tau) - \Psi^T(0) \int_{-h}^{0} e^{\lambda\tau} d\beta(\tau)$$

$$= B_\Lambda - \Psi^T(0) B(e^{\lambda\cdot}).$$

The next equation can be established by analogy by the use of the identity $\Psi(0)A_\Lambda^T = \dot{\Psi}(0) = L^T\Psi + M^T\dot{\Psi}$

$$(\lambda I - A_\Lambda) \langle \iota^T\Psi, Fe^{\lambda\cdot}\rangle = \Psi^T(0)\Delta(\lambda). \tag{64}$$

Now let $\det \Delta(\lambda) = 0$ and $\lambda \notin \Lambda$. Then there exists some nonzero $x \in \mathbb{C}^n$ such that $\Delta(\lambda)x = 0$. Hence $\phi = e^{\lambda\cdot}x \in \ker(\lambda I - A)$ (Lemma 2.4.1). Since $\lambda \notin \Lambda$, this implies $\phi \in X^\Lambda$ and thus $F\phi \perp \iota^T X_\Lambda^T$ (Theorem 2.4.6). In other words, $\langle \iota^T\Psi, Fe^{\lambda\cdot}\rangle\, x = 0$. By (62), we obtain $\Delta_K(\lambda)\binom{x}{0} = 0$ and hence $\det \Delta_K(\lambda) = 0$.

Secondly, let $\lambda \in \sigma(A_\Lambda + B_\Lambda K_\Lambda)$. Then there exists some nonzero $z \in \mathbb{C}^N$ such that $z^T(\lambda I - A_\Lambda) = z^T B_\Lambda K_\Lambda$. Defining

$$x := \Psi(0)z \in \mathbb{C}^n, \; u := B_\Lambda^T z \in \mathbb{C}^m,$$

we obtain by (64)

$$\begin{aligned} x^T\Delta(\lambda) &= z^T\Psi^T(0)\Delta(\lambda) \\ &= z^T(\lambda I - A_\Lambda) \langle \iota^T\Psi, Fe^{\lambda\cdot} \rangle \\ &= z^T B_\Lambda K_\Lambda \langle \iota^T\Psi, Fe^{\lambda\cdot} \rangle \\ &= u^T K_\Lambda \langle \iota^T\Psi, Fe^{\lambda\cdot} \rangle \end{aligned}$$

and by (63)

$$\begin{aligned} x^T B(e^{\lambda\cdot}) &= z^T\Psi^T(0)B(e^{\lambda\cdot}) \\ &= z^T\left(B_\Lambda - (\lambda I - A_\Lambda) \langle \iota^T\Psi, Ee^{\lambda\cdot} \rangle\right) \\ &= u^T\left(I - K_\Lambda \langle \iota^T\Psi, Ee^{\lambda\cdot} \rangle\right). \end{aligned}$$

Hence the row vector $(x^T \; u^T)$ is orthogonal to $\Delta_K(\lambda)$. Now suppose that $x = 0$ and $u = 0$. Then $z^T(\lambda I - A_\Lambda) = u^T K_\Lambda = 0$ and hence the following equation holds for $-h \leqslant \tau \leqslant 0$

$$\Psi(\tau)z = \Psi(0)e^{A_\Lambda^T\tau} z = e^{\lambda\tau}\Psi(0)z = e^{\lambda\tau}x = 0.$$

This means that $z = 0$, a contradiction. We conclude that $(x^T \; u^T) \neq 0$ and hence $\det \Delta_K(\lambda) = 0$.

Finally, let $\det \Delta_K(\lambda) = 0$ and $\lambda \notin \sigma(A_\Lambda + B_\Lambda K_\Lambda)$. Then there exists a non-zero pair $x \in \mathbb{C}^n$, $u \in \mathbb{C}^m$ such that

$$\Delta(\lambda)x = B(e^{\lambda\cdot})u, \tag{65.1}$$

$$K_\Lambda \langle \iota^T\Psi, Fe^{\lambda\cdot} \rangle x = u - K_\Lambda \langle \iota^T\Psi, Ee^{\lambda\cdot} \rangle u. \tag{65.2}$$

Defining $z := \langle \iota^T\Psi, Fe^{\lambda\cdot}x + Ee^{\lambda\cdot}u \rangle \in \mathbb{C}^N$, we obtain

$$\begin{aligned} B_\Lambda K_\Lambda z &= B_\Lambda u \\ &= \Psi^T(0)B(e^{\lambda\cdot})u + (\lambda I - A_\Lambda) \langle \iota^T\Psi, Ee^{\lambda\cdot}\rangle u \\ &= \Psi^T(0)\Delta(\lambda)x + (\lambda I - A_\Lambda) \langle \iota^T\Psi, Ee^{\lambda\cdot}\rangle u \\ &= (\lambda I - A_\Lambda)z. \end{aligned}$$

Here we have used equations (65.2), (63), (65.1) and (64). Now recall that $\lambda \notin \sigma(A_\Lambda + B_\Lambda K_\Lambda)$ and hence $z = 0$. By (65.2), this implies $u = K_\Lambda z = 0$ and thus $x \neq 0$. On the other hand, it follows from (65.1) that $\Delta(\lambda)x = B(e^{\lambda\cdot})u=0$. We conclude that $\det \Delta(\lambda) = 0$ and $\phi = e^{\lambda\cdot}x \in \ker(\lambda I - A)$. Finally, we have $\langle \iota^T\Psi, F\phi\rangle = \langle \iota^T\Psi, Fe^{\lambda\cdot}\rangle x = z = 0$ and hence $\phi \in X^\Lambda$. This shows that $\lambda \notin \Lambda$. □

As a consequence of Theorem 5.3.2, together with Lemma 5.3.1, we obtain the following criterion for stabilizability of NFDEs with a stable difference equation Σ_0.

5.3.3 COROLLARY. *Let* M *be given by* (3) *and suppose that the difference equation* Σ_0 *is stable with exponential decay rate* $\omega \leq 0$, *i.e.*, (54) *is satisfied for some* $\varepsilon > 0$. *Then the following statements are equivalent.*

(i) *There exists some* $K : M^p \to \mathbb{R}^m$ *and some* $\varepsilon > 0$ *such that* $\det \Delta_K(\lambda) \neq 0$ *for every* $\lambda \in \mathbb{C}$, $\operatorname{Re} \lambda \geq \omega - \varepsilon$.

(ii) *For every* $\lambda \in \mathbb{C}$, $\operatorname{Re} \lambda \geq \omega$, $\operatorname{rank}[\Delta(\lambda), B(e^{\lambda\cdot})] = n$.

SENSITIVITY

Let us now discuss the question of how the stability of the closed loop system reacts on small variations of the parameters.

If the difference equation Σ_0 remains unchanged, then it is relatively easy to see that the closed loop system remains stable after sufficiently small perturbations. This is a consequence of the following three facts which we will not prove.

1* If (54) is satisfied, then the complex function $\det \Delta_0(\lambda)$ is uniformly bounded away from zero on the right half-plane $\{\lambda \in \mathbb{C} | \operatorname{Re} \lambda \geq \omega\}$.

2* In the domain $\operatorname{Re} \lambda \geq 0$, the limit

$$\lim_{|\lambda|\to\infty} |\lambda^{-n} \det \Delta_K(\lambda) - \det \Delta_0(\lambda)| = 0$$

exists uniformly for bounded parameter values VAR η, VAR β, $\|A_{-\infty}\|_{W^{1,p}}$, $\|K\|_{L(M^p,\mathbb{R}^m)}$ (compare the proof of Lemma 5.2.4).

3* On every compact domain the zeros of det $\Delta_K(\lambda)$ depend continuously on the system and feedback parameters.

For the implementation of a dynamic observer, it is clearly necessary to allow (small) variations of the parameters h_j, A_{-j} of the difference equation. In view of the considerations in Section 5.1, we must assume that these variations do not affect the stability of Σ_0. Moreover, we need the stronger property that condition 1* is satisfied uniformly in the parameters h_j, A_{-j}. This is easy to check for systems with a single point delay, since in this case $\Delta_0(\lambda) = I - A_{-1}e^{-\lambda h}$.

We conclude that, for systems with a single point delay and a stable difference equation (*i.e.*, $|s| < e^{\omega h}$ $\forall$ $s \in \sigma(A_{-1})$), *the stability of the closed loop system is not affected by sufficiently small variations in all parameters.*

In general, this seems to be an open problem.

Finally, we will briefly point out the consequences of our results for the problem of stabilizing a NFDE by dynamic output feedback.

DYNAMIC COMPENSATION

Consider the NFDE

$$\Sigma \qquad \begin{aligned} \dot{w}(t) &= Lx_t + Bu_t \\ x(t) &= w(t) + Mx_t \\ y(t) &= \Gamma x_t \end{aligned}$$

with general delays in input, state and output. Let M be given by (3) and assume that the difference equation Σ_0 is stable, i.e., (54) holds for some $\varepsilon > 0$. Moreover, suppose that

$$\text{rank } [\Delta(\lambda) \;\; B(e^{\lambda \cdot})] = \text{rank} \begin{bmatrix} \Delta(\lambda) \\ \Gamma(e^{\lambda \cdot}) \end{bmatrix} = n \tag{66}$$

for every $\lambda \in \mathbb{C}$, Re $\lambda \geq \omega$. Then there exists a stable observer for system Σ, described by the equations

$$\begin{aligned} &\dot{\tilde{w}}(t) = L(\tilde{x}_t + H_1 * v_t) + H_0 v(t) + Bu_t \\ \Sigma_H \qquad &\tilde{x}(t) = \tilde{w}(t) + M(\tilde{x}_t + H_1 * v_t) \\ &v(t) = \Gamma(\tilde{x}_t + H_1 * v_t) - y(t) \end{aligned}$$

(Lemma 5.2.3, Theorem 5.2.7, Corollary 5.3.3). Moreover, there exists a stabilizing control law of the form (50), (52) (Theorem 5.2.11). In this control law we replace the state variables $w(t)$, $x(t)$ of system Σ by the state variables $\tilde{w}(t)$, $\tilde{x}(t)$ of the observer Σ_H. This leads to the following equation

$$\begin{aligned} u(t) = K(0)\tilde{w}(t) &+ \int_{-h}^{0}\int_{\tau}^{0} K(\tau-\sigma)d\eta(\tau)\tilde{x}(t+\sigma)d\sigma \\ &+ \int_{-h}^{0}\int_{\tau}^{0} \dot{K}(\tau-\sigma)d\mu(\tau)\tilde{x}(t+\sigma)d\sigma \qquad (67) \\ &+ \int_{-h}^{0}\int_{\tau}^{0} K(\tau-\sigma)d\beta(\tau)u(t+\sigma)d\sigma. \end{aligned}$$

It is easy to see that the 'error' variables $g(t) = \tilde{w}(t) - w(t)$, $e(t) = \tilde{x}(t) - x(t)$ together with $v(t)$ satisfy the homogeneous system Σ_H. Now replacing $\tilde{w}(t)$ and $\tilde{x}(t)$ in (67) by $w(t) + g(t)$ and $x(t) + e(t)$ shows that the closed-loop system Σ, Σ_H, (67) is stable with exponential decay rate ω.

Appendix

INPUT-OBSERVABILITY

In this section we consider the finite-dimensional linear system

$$\text{(A1)} \qquad \begin{aligned} \dot{x}(t) &= Ax(t) + Bu(t) \\ y(t) &= Cx(t) + Du(t) \end{aligned}$$

where $x \in \mathbb{R}^n$, $u \in \mathbb{R}^\ell$, $y \in \mathbb{R}^m$ and A, B, C, D are real matrices with the appropriate numbers of rows and columns.

A2 DEFINITION. *System* (A1) *is said to be input-observable if the following implication holds for every control function* $u \in L^p_{loc}(\mathbb{R},\mathbb{R}^\ell)$, $1 < p < \infty$,

$$\text{(A3)} \qquad y(t) \equiv 0, \ x(0) = 0 \Rightarrow x(t) \equiv 0.$$

A4 REMARK. If $D = 0$, then system (A1) is input-observable if and only if the maximal reachability subspace in ker C is zero. It has been proved in Moore-Laub [107] that this is equivalent to

$$\text{(A5)} \qquad \max_{\lambda \in \mathbb{C}} \operatorname{rank} \begin{bmatrix} A-\lambda I & B \\ C & 0 \end{bmatrix} = n + \operatorname{rank} B.$$

The following theorem generalizes this criterion to the case $D \neq 0$.

A6 THEOREM. *Let* $T_0 < T_1$ *be given. Then the following statements are equivalent.*

(i) *System* (A1) *is input-observable.*

(ii) *If* $y(t) = 0$ *for* $T_0 \leq t \leq T_1$ *and* $x(T_0) = 0$, *then* $x(t) = 0$ *for* $T_0 \leq t \leq T_1$.

(iii) *If* $y(t) = 0$ *for* $T_0 \leq t \leq T_1$ *and* $x(T_1) = 0$, *then* $x(t) = 0$ *for* $T_0 \leq t \leq T_1$.

(iv) *There exists some* $\lambda \in \mathbb{C}$ *such that*

(A7) $\quad \text{rank}\begin{bmatrix} A-\lambda I & B \\ C & D \end{bmatrix} = n + \text{rank}\begin{bmatrix} B \\ D \end{bmatrix}.$

A8 REMARKS

(i) Without loss of generality we can choose $T_0 = 0$ in the statements (ii) and (iii) of the above theorem, since system (A1) is time-invariant.

(ii) Condition (A7) implies that

$$\text{rank}\,[C\ D] \geq \text{rank}\begin{bmatrix} B \\ D \end{bmatrix}.$$

PROOF OF THEOREM A6. Without loss of generality we can assume that $\text{rank}\begin{bmatrix} B \\ D \end{bmatrix} = \ell$. Moreover, there exist unimodular matrices $M(\lambda)$ and $N(\lambda)$ such that

$$\text{(A9)} \qquad M(\lambda)\begin{bmatrix} A-\lambda I & B \\ C & D \end{bmatrix} N(\lambda) = \begin{bmatrix} \alpha_1(\lambda) & & 0 \\ & \ddots & \\ 0 & & \end{bmatrix}$$

is in Smith form. Then condition (iv) is satisfied iff all columns on the right-hand side of (A9) are nonzero.

Now suppose that (iv) does not hold. Then there exist nonzero polynomials

$$p(\lambda) = \sum_{j=0}^{k} p_j\lambda^j \in \mathbb{R}^n[\lambda], \quad q(\lambda) = \sum_{j=0}^{k} q_j\lambda^j \in \mathbb{R}^\ell[\lambda],$$

such that

$$(A - \lambda I)p(\lambda) + Bq(\lambda) = 0, \quad Cp(\lambda) + Dq(\lambda) = 0,$$

for all $\lambda \in \mathbb{C}$ or equivalently

$$\text{(A10)} \qquad \begin{aligned} &p_k = 0, \\ &p_{j-1} = Ap_j + Bq_j, \quad j = 1,\ldots,k, \\ &0 = Ap_0 + Bq_0, \\ &0 = Cp_j + Dq_j, \quad j = 0,\ldots,k. \end{aligned}$$

Moreover let $T \in \mathbb{R}$ and define

$$x(t) = \sum_{j=0}^{k} p_j \frac{(t-T)^{k-j}}{(k-j)!}, \quad u(t) = \sum_{j=0}^{k} q_j \frac{(t-T)^{k-j}}{(k-j)!}.$$

Then (A10) implies that $x(T) = 0$ and that the following equations hold for every $t \in \mathbb{R}$

$$\dot{x}(t) = Ax(t) + Bu(t), \; y(t) = Cx(t) + Du(t) = 0.$$

With appropriate values of T (= 0, T_0, T_1), this is a contradiction to (i), (ii), (iii), respectively.

Next we show that (iv) implies (iii) with $T_0 = 0$ and $T_1 = T$. For this purpose let $x(t)$ be a solution of (A1) such that $u(t) = 0$ and $x(t) = 0$ for $t \geqslant T$ and $y(t) = 0$ for $t \geqslant 0$. Then the Laplace transforms $\hat{x}(\lambda)$ and $\hat{u}(\lambda)$ of $x(t)$ and $u(t)$ ($t \geqslant 0$!) are entire functions satisfying

$$\text{(A11)} \quad \begin{bmatrix} A-\lambda I & B \\ C & D \end{bmatrix} \begin{pmatrix} \hat{x}(\lambda) \\ \hat{u}(\lambda) \end{pmatrix} = \begin{pmatrix} -x(0) \\ 0 \end{pmatrix}.$$

Since (iv) is satisfied, we have $m \geqslant \ell$ and can define $\tilde{M}(\lambda) \in \mathbb{R}^{(n+\ell)\times(n+m)}[\lambda]$ as consisting of the upper $n + \ell$ rows of $M(\lambda)$. Then it follows from (A9) that

$$\text{(A12)} \quad \tilde{M}(\lambda) \begin{bmatrix} A-\lambda I & B \\ C & D \end{bmatrix} N(\lambda) = \begin{bmatrix} \alpha_1(\lambda) & & 0 \\ & \ddots & \\ 0 & & \alpha_{n+\ell}(\lambda) \end{bmatrix}.$$

where all the $\alpha_j(\lambda)$ are nonzero polynomials. Combining (A11) and (A12), we obtain the following equation

$$\begin{pmatrix} \hat{x}(\lambda) \\ \hat{u}(\lambda) \end{pmatrix} = N(\lambda) \begin{bmatrix} \alpha_1(\lambda)^{-1} & & \\ & \ddots & \\ & & \alpha_{n+\ell}(\lambda)^{-1} \end{bmatrix} \tilde{M}(\lambda) \begin{pmatrix} -x(0) \\ 0 \end{pmatrix}.$$

This is an entire function (left-hand side) of exponential growth zero (right-hand side). Applying a theorem of Paley and Wiener (see, e.g., Rudin [131, Theorem 19.3]), we obtain that $x(t) = 0$ and $u(t) = 0$ for $t \geqslant 0$.

That (iv) implies (ii) follows from the fact that system (A1) has the property (ii) if and only if the (time inverse) system

$$\begin{aligned} \dot{x}(t) &= -Ax(t) - Bu(t) \\ y(t) &= Cx(t) + Du(t) \end{aligned}$$

has the property (iii). Finally, (i) follows trivially from (ii) and (iii). □

References

[1] H.T. Banks and J.A. Burns, Eigenmanifold decomposition for retarded functional differential equations in Hilbert space. Brown University, LCDS Technical Report TR-1, 1974.

[2] H.T. Banks and J.A. Burns, Hereditary control problems: numerical methods based on averaging approximations. SIAM J. Control Opt. 16 (1978), 169-208.

[3] H.T. Banks., M.Q. Jacobs and C.E. Langenhop, Applications of alternative methods to controllability of functional differential equations. "Optimal Control and its Applications", B.J. Kirby, ed., pp. 1-23, Springer-Verlag, New York, 1974.

[4] H.T. Banks., M.Q. Jacobs and C.E. Langenhop, Function space controllability of linear functional differential equations. "Differential Games and Control Theory", L. Sternberg, ed., Marcel Dekker, New York, 1974.

[5] H.T. Banks., M.Q. Jacobs and C.E. Langenhop, Characterization of the controlled states in $W_2^{(1)}$ of linear hereditary systems. SIAM J. Control 13 (1975), 611-649.

[6] H.T. Banks., M.Q. Jacobs and M.R. Latina, The synthesis of optimal controls for linear time optimal systems with retarded controls. J. Opt. Theory Appl. 8 (1971), 319-366.

[7] H.T. Banks and G. Kent, Control of functional differential equations to target sets in function space. SIAM J. Control 10 (1972), 567-593.

[8] H.T. Banks and A. Manitius, Projection series for retarded functional differential equations with applications to optimal control problems. J. Diff. Equations 18 (1975), 296-332.

[9] Z. Bartosiewicz, Density of images of semigroup operators for linear neutral functional differential equations. J. Diff. Equations 38 (1980), 161-175.

[10] Z. Bartosiewicz, Approximate controllability of neutral systems with delays in control. Ph.D. Dissertation, Institute of Mathematics, Polish Academy of Sciences, Warsaw, 1981.

[11] C. Bernier and A. Manitius, On semigroups in $\mathbb{R}^n \times L^p$ corresponding to differential equations with delays. Can. J. Math. 30 (1978), 897-914.

[12] K.P.M. Bhat and H.N. Koivo, An observer theory for time delay systems. IEEE Trans. Autom. Control, AC-21 (1976), 266-269.

[13] K.P.M. Bhat and H.N. Koivo, Modal characterization of controllability and observability for time delay systems. IEEE Trans. Autom. Control, AC-21 (1976), 292-293.

[14] C.W. Bitzer, Stieltjes-Volterra integral equations. Illinois J. Math. 14 (1970), 434-451.

[15] H. Bohr, Almost Periodic Functions. Chelsea Publ. Co., New York, 1947.

[16] J.G. Borisovic and A.S. Turbabin, On the Cauchy problem for linear nonhomogeneous differential equations with retarded argument. Soviet Math. Doklady 10 (1969), 401-405.

[17] J.A. Burns and T.L. Herdman, Adjoint semigroup theory for a class of functional differential equations. SIAM J. Math. Anal. 7 (1976), 729-745.

[18] J.A. Burns., T.L. Herdman and H.W. Stech, The Cauchy problem for linear functional differential equations. Proc. Conf. Integral and Functional Differential Equations, Morgantown, West Virginia, June 1979, T.L. Herdman, H.W. Stech, S.M. Rankin, eds., pp. 139-149, Marcel Deccer, New York, 1981.

[19] J.A. Burns., T.L. Herdman and H.W. Stech, Linear functional differential equations as semigroups in product spaces. Department of Mathematics, Virginia Polytechnic Institute and State University Blacksburg, Virginia, 1981. SIAM J. Math. Anal. 14 (1983), 98-116.

[20] L.A.V. Carvalho, An analysis of the characteristic equation of the scalar linear difference equation with two delays, in "Functional Differential Equations and Bifurcation", A.F. Izé, ed., pp. 69-81, Springer-Verlag, New York, 1980.

[21] A.K. Choudhury, A contribution to the controllability of time lag systems. Int. J. Control 17 (1973).

[22] D.H. Chyung, On the controllability of linear systems with delay in control. IEEE Trans. Autom. Control, AC-15 (1970), 255-257.

[23] C. Corduneanu, Almost Periodic Functions. Interscience, New York, 1968.

[24] R.F. Curtain and A.J. Pritchard, Infinite Dimensional Linear Systems Theory. Lecture Notes in Control and Information sciences, Vol. 8, Springer-Verlag, Berlin, 1978.

[25] R.F. Curtain and A.J. Pritchard, An abstract theory for unbounded control action for distributed parameter systems. Control Theory Centre, University of Warwick, Report No. 39, 1976.

[26] M.C. Delfour, State theory of linear, hereditary differential systems. J. Math. Anal. Appl. 60 (1977), 8-35.

[27] M.C. Delfour, The largest class of hereditary systems defining a C_o-semigroup on the product space. Can. J. Math. 32 (1980), 969-978.

[28] M.C. Delfour, Status of the state space theory of linear, hereditary differential systems with delays in state and control variables. "Analysis and Optimization of Systems", A. Bensoussan, J.L. Lions, eds., pp. 83-96, Springer-Verlag, New York, 1980.

[29] M.C. Delfour and A. Manitius, The structural operator F and its role in the theory of retarded systems. Part 1: J. Math. Anal. Appl. 73 (1980), 466-490, Part 2: J. Math. Anal. Appl. 74 (1980), 359-381.

[30] M.C. Delfour and S.K. Mitter, Hereditary differential systems with constant delays. Part 1: General case, J. Diff. Equations 12 (1972), 213-235, Part 2: A class of affine systems and the adjoint problem. J. Diff. Equations 18 (1975), 18-28.

[31] M.C. Delfour and S.K. Mitter, Controllability, observability, and optimal feedback control of affine hereditary differential systems. SIAM J. Control 10 (1972), 287-328.

[32] O. Diekmann, Volterra integral equations and semigroups of operators. Preprint, Mathematisch Centrum Report TW 197/80, Amsterdam 1980.

[33] O. Diekmann, A duality principle for delay equations. Preprint, Mathematisch Centrum Report TN 100/81, Amsterdam 1981.

[34] N. Dinculeanu, Integration on Locally Compact Spaces, Noordhoff, Leyden, 1974.

[35] S. Dolecki, Duality for various notions of controllability and observability. Proc. Int. Conf. on Diff. Equations., University of California, 1974.

[36] S. Dolecki and D.L. Russell, A general theory of observation and control.SIAM J. Control Opt. 15 (1977), 185-220.

[37] N. Dunford and J.T. Schwartz, Linear Operators, Part I: General Theory. Interscience, New York, 1957.

[38] R. Gabasov and M. Kirillova, Qualitative theory of optimal processes. Nauka, Moscow, 1971.

[39] J. Goldstein, Semigroups of Operators and Abstract Cauchy Problems. Lecture Notes, Tulane University, New Orleans, 1970.

[40] R.V. Gressang, Observers and pseudo observers for linear time delay systems. Proceedings of the 12th Allerton Conference on Circuit and Systems Theory, 1974.

[41] R.V. Gressang and G.B. Lamont, Observers for systems characterized by semigroups. IEEE Trans. Autom. Control, AC-20 (1975), 523-528.

[42] J.K. Hale, Theory of Functional Differential Equations. Springer-Verlag, New York, 1977.

[43] J.K. Hale and K.R. Meyer, A class of functional equations of neutral type. Mem. Amer. Math. Soc., No. 76, 1967.

[44] M.L.J. Hautus and E.D. Sontag, An approach to detectability and observers. Eindhoven University of Technology, Department of Mathematics, Report, 1980.

[45] M. Hazewinkel, A partial survey of the use of algebraic geometry in systems and control theory. Report 7913/M, Erasmus University, Rotterdam, 1979.

[46] D. Henry, Small solutions of linear autonomous functional differential equations. J. Diff. Equations 8 (1970), 494-501.

[47] D. Henry, The adjoint of a linear functional differential equation and boundary value problems. J. Diff. Equations 9 (1971), 55-66.

[48] D. Henry, Linear autonomous functional differential equations of neutral type in the Sobolev space $W_2^{(1)}$. Technical Report, Department of Mathematics, University of Kentucky, Lexington, Kentucky, 1970.

[49] D. Henry, Adjoint theory and boundary value problems for neutral linear functional differential equations. Preprint, Department of Mathematics, University of Kentucky, Lexington, Kentucky, 1970.

[50] D. Henry, Linear autonomous neutral functional differential equations. J. Diff. Equations 15 (1974), 106-128.

[51] G.A. Hewer and G.J. Nazaroff, Observer theory for delayed differential equations. Int. J. Control 18 (1973), 1-7.

[52] E. Hewitt and K.A. Ross, Abstract Harmonic Analysis I. Springer-Verlag, Berlin, 1970.

[53] E. Hille and R.S. Phillips, Functional Analysis and Semigroups. Amer. Math. Soc., Providence, R.I., 1957.

[54] D.B. Hinton, A Stieltjes-Volterra integral equation theory. Canad. J. Math. 18 (1966), 314-331.

[55] C.S. Hönig, Volterra Stieltjes-integral equations. North-Holland, Amsterdam, 1975.

[56] C.S. Hönig, Volterra Stieltjes-integral equations. "Functional Differential Equations and Bifurcation", A.F. Izé, ed., pp. 173-216, Springer-Verlag, New York, 1980.

[57] A. Ichikawa, Generation of semigroups on some product space with applications to evolution equations with delay. Control Theory Centre, University of Warwick, Report No. 52, 1976.

[58] A. Ichikawa, Optimal quadratic control and filtering for evolution equations with delay in control and observation. Control Theory Centre, University of Warwick, Report No. 53, 1976.

[59] K. Ito, Linear functional differential equations and control and estimation problems. Ph.D. Dissertation, Washington University, St. Louis, 1981.

[60] M.Q. Jacobs and C.E. Langenhop, Controllable two dimensional neutral systems. "Mathematical Control Theory", S. Dolecki, C. Olech, J. Zabczyk, eds., pp. 107-113, Banach Center Publications, Warsaw, 1976.

[61] M.Q. Jacobs and C.E. Langenhop, Criteria for function space controllability of linear neutral systems. SIAM J. Control Opt. 14 (1976), 1009-1048.

[62] B. Jakubczyk, A classification of attainable sets of linear differential-difference systems. Institute of Mathematics, Polish Academy of Sciences, Preprint 134, Warsaw, 1978.

[63] B. Jakubczyk and A.W. Olbrot, Dynamic feedback stabilization of linear time lag systems IVth IFAC Congress, Helsinki, 1978.

[64] B. Jakubczyk and A.W. Olbrot, Canonical delays, $\mathbb{R}^n(s)$ -, and function space controllability of linear differential-difference systems. Int. J. Control.

[65] E.W. Kamen, On an algebraic theory of systems defined by convolution operators. Math. Systems Theory 9 (1975), 57-74.

[66] E.W. Kamen, An operator theory of linear functional differential equations. J. Diff. Equations 27 (1978), 274-297.

[67] F. Kappel, Laplace transform methods and linear autonomous functional differential equations. Math. Institut, University of Graz, Bericht Nr. 64, 1976.

[68] T. Kato, Perturbation Theory of Linear Operators, Springer-Verlag, New York, 1966.

[69] F.M. Kirillova and S.V. Curakova, Relative controllability of systems with time lag.Dokl. Acad. Nauk USSR, 176 (1967), No. 6.

[70] J. Klamka, Relative controllability and minimum energy control of linear systems with distributed delays in the control IEEE Trans. Autom. Control, AC-21 (1976), 594-595.

[71] J. Klamka, On the controllability of linear systems with delays in the control. Int. J. Control 25 (1977), 875-883.

[72] J. Klamka, Observer for linear feedback control of systems with distributed delays in control and observation. Systems and Control Letters 1 (1982), 326-331.

[73] M. Kociecki, Final observability of time lag systems. Preprint, Warsaw, Poland, 1979.

[74] T.B. Kopeikina and V.V. Mulartchik, On observability and recognizability of systems with delays. Differential'nye Uravneniya 10 (1974), 933-936.

[75] A. Korytowski, Functional controllability of systems with delay.Archiwum Autom. i. Telem. 20 (1975), 19-28.

[76] N.N. Krasovskii, On the stabilizability of dynamic systems by supplementary forces. Differential'nye Uravneniya 1 (1965), 1-9.

[77] N.N. Krasovskii and A.B. Kurzhanskii, A contribution to the observability of time lag systems. Differential'nye Uravneniya 2 (1966), 299-308.

[78] N.N. Krasovskii and Y.S. Osipov, Stabilization of a controlled system with time delay. Engineering Cybernetics 1 (1963), 1-11.

[79] S. Kurcyusz and A.W. Olbrot, On the closure in $W^{1,q}$ of the attainable subspace of linear time lag systems. J. Diff. Equations 24 (1977), 29-50.

[80] W.H. Kwon and A.E. Pearson, Feedback stabilization of linear systems with delayed control. IEEE Trans. Autom. Control, AC-25 (1980), 266-269.

[81] R.H. Kwong, Structural properties and estimation of delay systems. Electronic Systems Laboratory, M.I.T., Report ESL-R-614, 1975.

[82] E.B. Lee, Linear hereditary control systems in "Calculus of Variations and Control Theory", pp. 47-72, Academic Press, New York, 1976.

[83] E.B. Lee and A.W. Olbrot, Observability and related structural results for linear hereditary systems. International Conference on Systems Engineering, University of Warwick, Coventry, 1980.

[84] E.B. Lee and A.W. Olbrot, On reachability over polynomial rings and a related genericity problem. Proc. 14th Conference on Informations Science and Systems, Princeton, N.J., 1980.

[85] N. Levinson and C. McCalla, Completeness and independence of the exponential solutions of some functional differential equations. Studies in Appl. Math. 53 (1974), 1-15.

[86] B.J. Lewin, Nullstellenverteilung ganzer Funktionen. Akademie-Verlag, Berlin, 1962.

[87] R.M. Lewis, Control delayed system properties via an ordinary model. Int. J. Control 30 (1979), 477-490.

[88] J.L. Lions, Optimal Control of Systems Governed by Partial Differential Equations. Springer-Verlag, Berlin, 1971.

[89] J.L. Lions and E. Magenes, Nonhomogeneous Boundary Value Problems and Applications. Vol. I - III, Springer-Verlag, Berlin, 1972-1973.

[90] A. Manitius, On controllability conditions for systems with distributed delays in state and control. Archiwun Autom. i. Telem. 17 (1972), 363-377.

[91] A. Manitius, Controllability, observability, and stabilizability of retarded systems. Proc. 1976 IEEE Conf. on Decision and Control, pp. 752-758, IEEE Publications, New York, 1976.

[92] A. Manitius, Function space controllability of retarded systems: some new algebraic conditions. Centre de Recherches Mathématiques, Université de Montréal, CRM-Report No. 656, 1976.

[93] A. Manitius, Completeness and F-completeness of eigenfunctions associated with retarded functional differential equations. J. Diff. Equations 35 (1980), 1-29.

[94] A. Manitius, Necessary and sufficient conditions of approximate controllability for general linear retarded systems. SIAM J. Control Opt. 19 (1981), 516-632.

[95] A. Manitius, F-controllability and observability of linear retarded systems. Applied Math. Opt., 9 (1982), 73-95.

[96] A. Manitius and A.W. Olbrot, Controllability conditions for linear systems with delayed state and control. Arch. Autom. i. Telem. 17 (1972), 119-131.

[97] A. Manitius and A.W. Olbrot, Finite spectrum assignment problem for systems with delays IEEE Trans. Autom. Control, AC-24 (1979), 541-553.

[98] A. Manitius and R. Triggiani, Function space controllability of linear retarded systems: a derivation from abstract operator conditions. SIAM J. Control Opt. 16 (1978), 599-645.

[99] A. Manitius and R. Triggiani, New results on functional controllability of time delay systems. Proc. 1976 Conf. on Information Sciences and Systems, pp. 401-405, John Hopkins University, Baltimore, Maryland, 1976.

[100] A. Manitius and R. Triggiani, Sufficient conditions for function space controllability and feedback stabilizability of linear retarded systems. IEEE Trans. Autom. Control, AC-23 (1978), 659-664.

[101] V.M. Marchenko, On complete controllability of systems with delay·Probl. of Control and Inf. Theory 8 (1979), 421-432.

[102] A.I. Markushevich, Theory of Functions of a Complex Variable. Chelsea Publ. Co., New York, 1965.

[103] W.R. Melvin, Stability properties of functional differential equations. J. Math. Anal. Appl. 48 (1974), 749-763.

[104] R.K. Miller, Linear Volterra integro-differential equations as semi-groups. Funkcial. Ekvac. 17 (1974), 39-55.

[105] S.A. Minjuk, On complete controllability of linear controllable systems with delay. Differential'nye Uravneniya 8 (1972), 254-259.

[106] S.A. Minjuk and N.N. Stepanjuk, The theory of completely controllable linear systems with delay. Differential'nye Uravneniya 10 (1974), 629-634.

[107] B.C. Moore and A.J. Laub, Computation of supremal (A,B)-invariant and controllability subspaces. IEEE Trans. Autom. Control, AC-23 (1978), 783-792.

[108] A.S. Morse, Ring models for delay differential systems. Automatica 12 (1976), 529-531.

[109] D.A. O'Connor, State controllability and observability for linear neutral systems. Ph.D. Dissertation, Washington University, St. Louis, 1978.

[110] D.A. O'Connor and T.J. Tarn, On the function space controllability of linear neutral systems. SIAM J. Control Op. 21 (1983) 306-329.

[111] D.A. O'Connor and T.J. Tarn, On the stabilization by state feedback for neutral differential-difference equations. Department of Systems Science and Mathematics, Washington University, St. Louis, 1981.

[112] A.W. Olbrot, On controllability of linear systems with time delays in control. IEEE Trans. Autom. Control, AC-17 (1972), 664-666.

[113] A.W. Olbrot, On the existence of a stabilizing control of linear systems with delays in control. Archiwum Autom. i. Telem. 17 (1972).

[114] A.W. Olbrot, Algebraic criteria of controllability to zero function for linear constant time lag systems. Control and Cybernetics 2 (1973), 59-77.

[115] A.W. Olbrot, Observability tests for linear constant time lag systems. Control and Cybernetics 4 (1975), 71-84.

[116] A.W. Olbrot, A counterexample to "observability of linear systems with time variable delays". IEEE Trans. Autom. Control, AC-22 (1977), 821-283.

[117] A.W. Olbrot, Control of retarded systems with function space constraints Part 2: Approximate controllability. Control and Cybernetics 6 (1977), 17-69.

[118] A.W. Olbrot, Stabilizability, detectability, and spectrum assignment for linear systems with general time delays. IEEE Trans. Autom. Control, AC-23 (1978), 887-890.

[119] A.W. Olbrot, Observability and observers for a class of linear systems with delays, IEEE Trans. Autom. Control.

[120] A.W. Olbrot and A. Gasiewski, The effects of feedback delays on the performance of multivariable linear control systems. IEEE Trans. Autom. Control, AC-25 (1980).

[121] A.W. Olbrot and S.H. Zak, Sufficient and necessary conditions for spectral controllability and observability of linear retarded systems. Bull. Polon. Acad. Sci. Ser. Techn.

[122] A.W. Olbrot and S.H. Zak, Controllability and observability problems for linear functional differential systems. Foundations of Control Engineering 5 (1980), 79-89.

[123] Y.S. Osipov, Stabilization of controlled systems with delays. Differential'nye Uravneniya 1 (1965), 605-618.

[124] L. Pandolfi, On the infinite dimensional controllability of differential control processes. Bollettino U.M.I. 10 (1974), 114-123.

[125] L. Pandolfi, Feedback stabilization of functional differential equations. Bollettino U.M.I. 12 (1975), 626-635.

[126] L. Pandolfi, Stabilization of neutral functional differential equations. J. Opt. Theory Appl. 20 (1976), 191-204.

[127] A. Pazy, Semigroups of Linear Operators and Applications to Partial Differential Equations. Lecture Notes No. 10, University of Maryland, College Park, 1974.

[128] J. Pollock and A.J. Pritchard, The linear quadratic cost control problem for distributed systems with unbounded control action. Control Theory Centre, University of Warwick, Report No. 76, 1979.

[129] V.M. Popov, On the property of reachability for some delay differential equations. Technical Report R-70-08, University of Maryland, College Park, 1970.

[130] H.R. Rodas and C.E. Langenhop, A sufficient condition for function space controllability of a linear neutral system. SIAM J. Control Opt. 16 (1978), 429-435.

[131] W. Rudin, Real and Complex Analysis. McGraw-Hill, New York, 1970.

[132] D. Salamon, Observers and duality between observation and state feedback for time delay systems. IEEE Trans. Autom. Control, AC-25 (1980), 1187-1192.

[133] D. Salamon, On finite dimensional perturbation of semigroups corresponding to differential delay systems. Forschungsschwerpunkt Dynamische Systeme, Universität Bremen, Report Nr. 14, 1980.

[134] D. Salamon, On controllability and observability of time delay systems Forschungsschwerpunkt Dynamische Systeme, Universität Bremen, Report Nr. 38, 1981. IEEE Trans. Autom. Control, to appear.

[135] D. Salamon, On dynamic observation and state feedback for time delay systems, in "Conference on Differential Equations and Delays", June 1981, F. Kappel, W. Schappacher, eds., Research Notes in Mathematics, Pitman, London, 1982.

[136] J.M. Schumacher, Dynamic feedback in finite- and infinite-dimensional linear systems. Doctoral Dissertation, Mathematisch Centrum Amsterdam, 1981.

[137] S. Schwabik, On Volterra-Stieltjes integral equations. Cas. pro Pest. Mat. 99 (1974), 255-278.

[138] S. Schwabik, Note on Volterra-Stieltjes integral equations. Cas. pro Pest. Mat. 102 (1977), 275-279.

[139] O. Sebakhy and M.M. Bayoumi, A simplified criterion for the controllability of a linear system with delays in control. IEEE Trans. Autom. Control, AC-16 (1971), 364-365.

[140] S.N. Shimanov, On the theory of linear differential equations with after effect. Differential'nye Uravneniya 1 (1965), 102-116.

[141] E.D. Sontag, Linear systems over commutative rings: a survey. Ric. di Automatica 7 (1976), 1-34.

[142] A.E. Taylor, Introduction to Functional Analysis, Wiley & Sons, New York, 1958.

[143] R. Triggiani, Extensions of rank conditions for controllability and observability to Banach spaces and unbounded operators. SIAM J. Control 14 (1976), 313-338.

[144] R. Triggiani and A.J. Pritchard, Stabilizability in Banach spaces. Control Theory Centre, University of Warwick, Report No. 35, 1976.

[145] R.B. Vinter, On the evolution of the state of linear differential equations in M^2. J. Inst. Maths. Applics. 21 (1978), 13-23.

[146] R.B. Vinter, Semigroups on product spaces with applications to initial value problems with nonlocal boundary conditions. "Control of Distributed Parameter Systems", S.P. Banks, A.J. Pritchard, eds., pp. 91-98, Pergamon Press, Oxford, 1978.

[147] R.B. Vinter and R.H. Kwong, The infinite time quadratic control problem for linear systems with state and control delays: an evolution equation approach. SIAM J. Control Opt. 19 (1981), 139-153.

[148] K. Watanabe and M. Ito, An observer for linear feedback control of multivariable systems with multiple delays in control and output. Systems & Control Letters 1 (1981), 54-59.

[149] J. Zabczyk, Remarks on the algebraic Riccati equation in Hilbert space. Appl. Math. Anal. Opt. 2 (1976), 251-258.

[150] V. Zakian and N.S. Williams, A ring of delay operators with applications to delay differential systems. SIAM J. Control Opt. 15 (1977), 247-255.

[151] R.B. Zmood, On the Euklidean space and function space controllability of control systems with delay. Doctoral Dissertation, University of Michigan, 1971.

[152] R.B. Zmood, The Euklidean space controllability of control systems with delay. SIAM J. Control 12 (1974), 609-623.